# Guide to the International Marine Engineering Equipment Science and Technology Innovation Competition (2019-2022)

# 国际海洋工程装备科技创新大赛指南
## （2019—2022）

宋大雷　方奇志　主编

中国海洋大学出版社
CHINA OCEAN UNIVERSITY PRESS
·青岛·

图书在版编目（CIP）数据

国际海洋工程装备科技创新大赛指南 : 2019—2022 / 宋大雷, 方奇志主编. — 青岛 : 中国海洋大学出版社, 2022.7

ISBN 978-7-5670-2707-7

Ⅰ. ①国… Ⅱ. ①宋… ②方… Ⅲ. ①海洋工程—工程设备—科学技术—竞赛—世界—指南—2019-2022 Ⅳ. ①P75-62

中国版本图书馆CIP数据核字(2020)第257806号

出版发行　中国海洋大学出版社
社　　址　青岛市香港东路 23 号　　　　邮政编码　266071
出 版 人　刘文菁
网　　址　http: //pub.ouc.edu.cn
订购电话　0532-82032573（传真）
责任编辑　滕俊平
照　　排　青岛光合时代文化传媒有限公司
印　　制　青岛海蓝印刷有限责任公司
版　　次　2022 年 9 月第 1 版
印　　次　2022 年 9 月第 1 次印刷
成品尺寸　210 mm × 285 mm
印　　张　9.5
印　　数　1~2000
字　　数　170 千
定　　价　78.00 元

国际海洋工程装备科技创新大赛指南（2019—2022）

## 编委会

# 序言

众所周知，海洋是生命的摇篮、资源的宝库、交通的要道、气候的调节器，是人类解决资源短缺问题、拓展生存发展空间的战略要地，海洋对人类的生存和发展至关重要。我国拥有300多万平方千米的“蓝色国土”，是名副其实的海洋大国，也是世界上最早利用海洋的国家之一。战国时期，姜太公治齐倡鱼盐之利；早在秦汉时期，海上丝绸之路雏形便已形成；宋元时期，指南针的改进和造船技术的发展推动了航海技术进步；明朝前期，郑和下西洋，促进了经济文化交流。21世纪，海洋再度成为世界关注的焦点。顺应时代潮流，2012年，党的十八大报告提出要建设海洋强国，海洋的地位被提升到前所未有的战略高度。

海洋科技与工程装备是认识海洋、开发利用海洋以及保护海洋生态健康必要的支撑与基石。从世界范围来看，主要海洋国家高度重视海洋工程技术与装备的研发，发布了一系列科技发展计划、中长期发展规划，支持高端海洋科考船、海洋观测设备、海洋探测平台以及海上和水下生产作业装备的研发。经过多年发展，我国近浅海工程技术与装备设计建造和运维能力显著增强，并逐步转向“深海探测”“深海开发”的新阶段。例如，我国在载人潜水器、无人潜水器等方面的研究和应用取得较大进展，部分技术达到世界先进水平。2020年11月，载人潜水器“奋斗者”号在马里亚纳海沟多次完成万米级下潜，标志着我国具有了进入世界海洋最深处开展科学探索和研究的能力，体现了我国在海洋高技术领域的综合实力；海洋科考船队向综合性、现代化、大型化、谱系化方向发展，形成了“向阳红”系列、“雪龙”系列、“东方红”系列、“大洋”系列科考船队，基本满足近海、中海、远海、南北极的海洋科考需求。这些海洋工程技术与装备的进步，增强了我们问海、向海的底气和勇气。

中国海洋大学作为中国世界一流大学建设高校中唯一一所综合性海洋大学，因海而兴、向海而强，承担着推动海洋科技交流、海洋科技成果转化的职责和使命。从2019年起，中国海洋大学牵头主办了国际海洋工程装备科技创新大赛，不仅提供竞技交流的平台，更旨在培养一批有技术、有创意、能创新的海洋科技后备人才。经过两届竞赛，专家评委们评选出了一批有创意、高质量的大赛作品，为此，我们将部分获奖作品结集成册。这其中，既有对海洋世界的幻想描绘，也有解决海洋难题的实践方案。如在作品《摩西教授的海洋之约》中，参赛的小学生脑洞大开，通过搭载“仿生章鱼”，在未知的海洋中找到失踪已久的摩西教授。在这一过程中，他们不仅描绘了对未来仿生技术、海洋探测技术、导航技术等的应用设想，也生动展现了海底的旖旎风光。在作品《多航态三栖变体航行器》中，参赛大学生设计了一款跨介质航行器，对机械结构、动力系统、新材料使用等进行了细致的考虑，展示了良好的工程设计能力。我们希望，通过展示这些优秀作品，进一步浓厚全社会关注海洋、认识海洋、了解海洋、热爱海洋、保护海洋、开发海洋、经略海洋的良好氛围，更期待越来越多的人投入海洋工程科技与装备的学习、探索、研发中。

在本书的编纂过程中，诸多高等院校、科研院所的专家同行提供了建设性的意见和建议，并为获奖作品写下了中肯、有指导性的评语；主办、承办比赛的各个单位、工作人员，在办赛、出版筹备的过程中，承担了许多烦琐的工作。正是大家的共同努力，促使本书成功面世，我想这也是我们为海洋人才培养、海洋强国发展做出的实实在在的贡献。

李华军

2022年3月于青岛

# PREFACE

As we all know, the ocean is the cradle of lives, the treasure house of resources, the bottle neck of traffics, the regulator of the climate, the solution to shortage of resources and a strategic space to be expanded for survival and development. Our nation has more than 3 million square kilometers of "blue land" and China is a major maritime country and one of the first group of countries in the world to make use of the ocean. During the Warring States Period, Jiang Taigong ruled the ancient state of Qi and advocated the benefits of fishing and salt; as early as the ancient Qin and Han Dynasties, the prototype of the Maritime Silk Road had been formed; during the ancient Song and Yuan Dynasties, with the improvement of the compass and the development of shipbuilding technology, the advancement of navigation technology was promoted; as early as Ming Dynasty, Zheng He's Expeditions to the West promoted economic and cultural exchanges and left many geographical works to the world. In 21st century, the ocean has once again become the focus of the world. Conforming to the trend of times, in 2012, the report of the 18th CPC National Congress proposed to build a strong maritime nation, and the ocean has been raised to an unprecedented strategic level.

It is the necessary support and cornerstone for understanding, developing, utilizing the ocean and maintaining a healthy environment of the ocean by marine technology and engineering equipment. From the perspective of the world, major marine countries have attached great importance to the research and development of marine engineering technology and equipment, and have issued a series of scientific and technological development plans, medium and long-term development plans, they have supported the research and development of high-end marine scientific research vessels, marine observation equipment, marine exploration platforms, and offshore and underwater production and operation equipment. After years of development, the design, construction, operation and maintenance capabilities of China's near shallow water engineering technology and equipment have been significantly increased, and gradually turned to a new stage of "deep-sea exploration" and "deep-sea development". For example, China has made great progress in the research and application of HOV, UUV, and underwater gliders, and some technologies have reached the world-advanced standards. In November 2020, the Striver completed several 10,000-meter-depth of dives in the Mariana Trench, marking that our nation has the ability to enter the deepest part of the world's ocean to carry out scientific exploration and research, and reflects our nation's comprehensive strength in the field of marine high technology. The marine scientific research fleets have developed in the direction of comprehensiveness, modernization, large-scale and pedigree, forming the "Xiangyanghong" series, "Xuelong"series, "Dongfanghong" series and "Dayang" series scientific research fleets, which basically meet the needs of marine scientific research in offshore, middle sea, open sea, the Antarctic and the Arctic, etc. These advances in marine engineering equipment and technology have continuously strengthened our confidence and courage to develop the ocean.

Ocean University of China, as the only comprehensive marine university among the class A level of the world-class universities in China, it is thrived and developed because of the ocean, and it undertakes the responsibilities and missions of promoting the exchange of marine sciences and technologies and the transformation of scientific and technological achievements. Since 2019, Ocean University of China has taken the lead in hosting the International Marine Engineering Equipment Technology Innovation Competition, which not only provides a platform for competitive communications, but also aims to cultivate a group of marine science and technology reserve talents with technology, creativity and innovation. After two competitions, the expert judges selected a number of creative and high-quality competition works. Therefore, we have collected some of the award-winning works into a book, including both the fantasy descriptions of the marine world and the practical solutions to solve marine problems. For example, in the work of *Professor Munsey's Ocean Covenant*, the participating primary school students opened their minds and found long-lost Professor Munsey in the unknown ocean area by carrying a "Bionic Octopus". In the process, they did not only describe the application of bionic technology, marine exploration technology, and navigation technology in the future, but also vividly displayed the beautiful scenery of the seabed. In the work of "Multi-state Variant Triphibious Vehicle", the participating college student designed a cross-media vehicle, this work carefully considered the mechanical structure, power system, the use of new materials, etc., and demonstrated great engineering design capabilities. We hope that by presenting these excellent works, we could further establish the good atmosphere of the whole society paying attention to the ocean, knowing the ocean, understanding the ocean, loving the ocean, protecting the ocean, developing the ocean, and managing the ocean, and we expect more and more people to devote themselves into study, exploration and research and development of marine engineering technology and equipment.

In the process of compiling this collection of works, experts from many colleges and research institutes have provided constructive opinions and suggestions, and have written down pertinent and instructive comments on the award-winning works; and in the meanwhile, all units and staffs hosting or undertaking the competition have undertaken a lot of cumbersome works in the process of organizing the competition and publishing preparation. Because of the joint efforts of all of you that this collection of works has been successfully published, and from my perspective, this is also our real contribution to the cultivation of marine talents and the development of a strong maritime nation.

Huajun Li

March, 2022, in Qingdao

# 目录 | CONTENTS

## 首届国际海洋工程装备科技创新大赛

The First International Marine Engineering Equipment Science and Technology Innovation Competition

## 第二届国际海洋工程装备科技创新大赛

## The 2nd International Marine Engineering Equipment Science and Technology Innovation Competition



## 第三届国际海洋工程装备科技创新大赛

## The 3rd International Marine Engineering Equipment Science and Technology Innovation Competition

# 首届国际海洋
# 工程装备科技创新大赛

The First International Marine
Engineering Equipment Science and
Technology Innovation Competition

# 在首届国际海洋工程装备科技创新大赛颁奖典礼上的致辞

中国海洋大学党委常委、副校长　李巍然

尊敬的各位嘉宾、各位专家，老师们、同学们：

大家上午好！今天，我们相聚在这里，隆重举行首届国际海洋工程装备科技创新大赛颁奖典礼。作为大赛的主办单位，我谨代表中国海洋大学向各位获奖者表示热烈的祝贺，向关心、支持大赛的社会各界和励展博览集团表示衷心的感谢！

大赛自2018年10月启动以来，得到了国内外高校、科研院所、涉海企业的广泛关注。初赛阶段，156支队伍、536名参赛选手提交了参赛作品。经过网上初评，有107支队伍、300余名选手参加了11月9—10日在中国海洋大学举办的大赛决赛，评选出了一批创意新颖、质量上乘的大赛作品，涌现出一批有思想、懂技术、能创新的海洋科技后备人才。

海洋是人类发展的第二大空间，是地球上远未充分开发的资源宝库。保护海洋生态、开发海洋资源、利用海洋空间、发展海洋经济，已经成为世界多数国家的发展战略。海洋工程装备作为发展海洋经济、海洋科技的重要支撑，在人类探索开发海洋、保护海洋的事业中具有举足轻重的作用。在海洋工程装备领域互通新需求、交流新技术、发布新产品、强化新合作正在成为国际社会的新潮流。

中国海洋大学作为中国世界一流大学建设高校中唯一一所综合性海洋大学，因海而兴、向海而强，承担着推动国际海洋科技交流、海洋科技成果转化的职责和使命。励展博览集团自2013年起主办了五届水下机器人大赛，通过组织大赛，培养了海洋科技人才、催生了海洋科技企业、促进了海洋事业发展。这次我们学校和企业联合主办大赛，不仅丰富了参赛作品的种类，也扩大了参赛人群的范围。广大青少年学生参加大赛，为大赛增加了活力，让海洋科技创新充满生机。我们国家关心海洋、认识海洋、经略海洋的队伍又多了一条延续血脉。我们完全可以期待，今天参加大赛的青少年学生，在未来的海洋强国建设中，一定会能担当、有作为、立功勋。我们也完全可以相信，只要社会各界、国内外同携手，共建海洋科技创新交流平台，就一定能推动全球海洋事业健康发展、美好发展。

衷心希望社会各界继续关心国际海洋工程装备科技创新大赛，继续支持国际海洋工程装备科技创新大赛，让我们共同努力，把大赛办成一个海洋未来人才的集结地、一个海洋工程装备的技术库、一个海洋科技创新的成果仓。

再次感谢各位嘉宾出席今天的颁奖典礼，同时预祝第七届Oi China上海国际海洋技术与工程装备展览会暨无人航行器大会圆满成功。

谢谢大家！

2019年11月13日

# Speech to the Award Ceremony of the First International Marine Engineering Equipment Science and Technology Innovation Competition

Member of the Standing Committee and Vice President of Ocean University of China, Mr. Li Weiran

---

Ladies and Gentlemen:

Good morning everyone! Today, we are gathering here to hold the award ceremony of the First International Marine Engineering Equipment Science and Technology Innovation Competition. As the organizer of the competition, on behalf of Ocean University of China, I'd like to give my warmest congratulations to all the winners, and sincerely thank all of you who care about and support the competition and heartily appreciate Reed Exhibitions!

Since its start in October 2018, the competition has received widespread attention from domestic and foreign universities, research institutes and maritime companies. In the preliminary stage, 536 competitors in 156 teams submitted their entries. More than 300 players in 107 teams passed the preliminary online evaluation and participated in the final of the competition held in Ocean University of China from November 9th to 10th, and selected lots of innovative and high-quality competition entries, and at the same time lots of marine science and technology backup talents who are ideological, understanding the technology, and able to innovate emerged.

Ocean is the second largest space for human development and a great treasure of resources that is not fully developed on the earth. It has become the development strategy by lots of countries in the world to protect marine ecology, develop marine resources, utilize marine space and develop marine economy. As a critical support for the development of marine economy and marine science & technology, marine engineering equipment plays an important role in the cause of human exploration, development and protection of the ocean. It has become a new trend for the international community to share with each other the new needs, exchange new technologies, release new products and strengthen new cooperation in the field of marine engineering equipment.

Ocean University of China, as the only comprehensive marine university in China's world-class university, is being prosperous and strong because of the sea. And OUC undertakes the responsibility and mission for promoting international marine science and technology exchange, and transformation of scientific and technological achievements. Since 2013, Reed Exhibitions has hosted five times underwater robot competitions. By organizing the competitions, it has developed marine science and technology talents, spawned marine science and technology enterprises, and promoted the development of marine businesses. This time, the university and the enterprise jointly hosted the competition, which not only expanded the types of entries, but also expanded the range of the participants. Lots of young students participate in the competition, which provides vitality to the competition and the marine science and technology innovation. So that there is another blood line which cares about, understands and manages the ocean in our nation. We can totally expect that the young students who participate in today's competition will certainly be able to take the responsibilities, make achievements and make contributions in the future construction of a powerful marine nation. We can also fully believe that as long as all communities, domestic and international counterparts work together to build a platform for marine science and technology innovation to share the information, it will surely promote the development of the global marine business.

I sincerely hope that everyone in our community could continue to care about the Marine Engineering Equipment Science and Technology Innovation Competition and continue to support it. Let's work together to make the competition a gathering place for future marine talents, a technology base for marine engineering equipment and an achievement warehouse for marine science and technology innovation.

Thank you again for attending today's award ceremony and wish the 7th Oi China Shanghai International Marine Technology and Engineering Equipment Exhibition and Unmanned Underwater Vehicle Conference a complete success.

Thank you all!

November 13th, 2019

**大赛主题：海洋时空下的航行器**

人类的时空认知主要来源于生活生产经验。日出日落有了天的概念；月亮的圆缺有了月的概念，也有了古代圭表、日晷的发明。并且，人类利用星星来辨别方向，用地理的标志来定位……然而，海洋时空与陆地有着本质的区别，尤其是在深海中，不仅看不到太阳、月亮和星星，就连定位的标志物也没有。海洋时空介质海水的流动，更让海洋这个世界变得神秘莫测。虽然科技在进步和发展，但是在海洋深处依然没有手机信号，也无法用GPS定位。海洋时空下的航行器，其设计理念和实现方式，与我们身边的运载工具有着太多的区别。本届大赛将以海洋时空下的航行器为主题，开展海洋工程装备的创意、设计、制作，完成工程任务，进行竞技。

Theme: Vehicles in Space-Time of Ocean

Human's knowledge of time and space mainly comes from their life. There are sunrise and sunset, they get the idea of day; the changes of moon in shape enlightens people; the stars are used for direction, and the geographical markers are utilized for positioning. And same goes for the time and space appliances, such as the Gui, the sundial and the general nautical map in ancient times. However, there is large variation between the ocean and the land in time and space, especially in the center of the ocean, where it is impossible to see the sun, moon and the stars, even the markers for positioning. The flow of seawater makes the world of the ocean mysterious and incredible. Although technology keeps on changing, there is still no signal and thus it is unavailable to GPS positioning in the deep sea. When it comes to the marine vehicles, it is quite different from the ordinary vehicles around us in its design and implementation. This competition will focus on their design and production of marine engineering equipment with aim to complete engineering tasks.

# 科幻类竞赛规则及特等奖作品

Competition Rules and Grand Prize Works
in Science Fiction Category

### 1.比赛题目

想象在深海环境中组织一场深海聚会：在同一时刻、同一地点，至少有两人乘坐海洋航行器在深海相聚。构想一种计时、定位的方法和装备，能够准确地进行深海计时与定位，使大家能够准时抵达约定地点。同时，基于上述计时、定位的方法和设备，构想一种航行器，以运载大家完成深海聚会。

### 2.比赛形式

科幻绘本。

### 3.比赛场地

预赛在网上进行，决赛在中国海洋大学进行。

### 4.比赛说明

1）参赛对象

社会组：已毕业在职人员；

大学组：在校研究生、本科生、专科生；

中小学组：在校高中生、初中生、小学生；

每组队员不超过4人。

2）参赛作品要求

围绕大赛主题，以深海聚会为故事背景提交绘本作品，形式为手绘或计算机绘图，字数、篇幅不限，上传文件为PDF格式，文件大小不超过5M，现场提交作品以A4纸打印。

3）比赛流程

比赛分为预赛和决赛两个阶段。

预赛阶段，采取专家网评形式，参赛者提交作品电子版（手绘为扫描版），作品择优进入决赛。

决赛阶段，入围作品进行路演，每支队伍路演时间不超过5分钟。

### 5.比赛规则

比赛采取评分形式，得分高者获胜。

满分为100分，具体评分标准如下。

（1）要素的准确性（30分）：绘本中包含聚会用海洋航行器（0分或5分）、计时器（0分或5分）、定位装置（0分或5分）、至少有两个人物（0分或5分）、聚会时间（0分或5分）、聚会地点（0分或5分）。

（2）作品的创意（50分）：故事情节的整体创意（1~5分）、海洋航行器的创意（1~15分）、计时器的创意（1~15分）、定位装置的创意（1~15分）。

（3）作品的文学性（1~10分）。

（4）作品的艺术性（1~10分）。

对专家的评分进行统计，去掉一个最高分，去掉一个最低分，其余评分取平均值，作为该作品的最终得分。

## 1.Description

In the deep ocean, we image a party when the marine lives meet at the same time and the same place. Please design a timing method and equipment that allows them to be independently timed, but to reach the place at the same time; design a positioning method and equipment that allows the two marine organisms to arrive at the same place; design a type of vehicles that carries them to complete their party based on the methods and equipment mentioned above.

## 2.Form

To complete a sci-fi picture book that is aimed at them.

## 3.Venue

The intermediary contest will be held online and the finals will be held at Ocean University of China.

## 4.More Information

1) Participants

Social group: working individuals;

University group: graduate students, undergraduates, and junior college students;

Primary school and high school group: primary school and high school students.

There are no more than 4 players in each group.

2) Requirements for works

To complete a science fiction book around the story of party in the deep ocean. The work should be less than 5 M, PDF format, the form of drawings by hand or computer.

3) Procedure

The competition is made up of the intermediary contest and finals.

For the intermediary contest, experts will be invited to select online. Participants are required to submit electronic versions of the works (those by hand should be scanning), and quality-based works will be netted.

With regarding to finals, the finalists will perform roadshows, and each will be given no more than 5 minutes.

## 5.Rules

Scoring mechanism will be used.

The score is 100 points and scoring rules are as follow:

(1) Elements (30 points): the picture book includes marine vehicles (0 or 5 points), timer (0 or 5 points), positioning device (0 or 5 points), more than 2 marine creatures (0 or 5 points), party time (0 or 5 points), party place (0 or 5 points).

(2) Creation ( 50 points): the idea of the story (1 to 5 points), the idea of the timer (1 to 15 points), the idea of the positioning device (1 to 15 points) and the idea of the dating aircraft (1 to 15 points).

(3) Literature (1 to 10 points).

(4) Artistic (1 to 10 points).

The highest score and the lowest score will be removed, then the total score of the referee and the average value will be counted as the score of the work.

## 《梦想巡航》大连海洋大学

*Dream Cruise*
Dalian Ocean University


神锋队

王泽明　王洋　李子涵　周瑞格

Shenfeng Team
Wang Zeming, Wang Yang, Li Zihan, Zhou Ruige

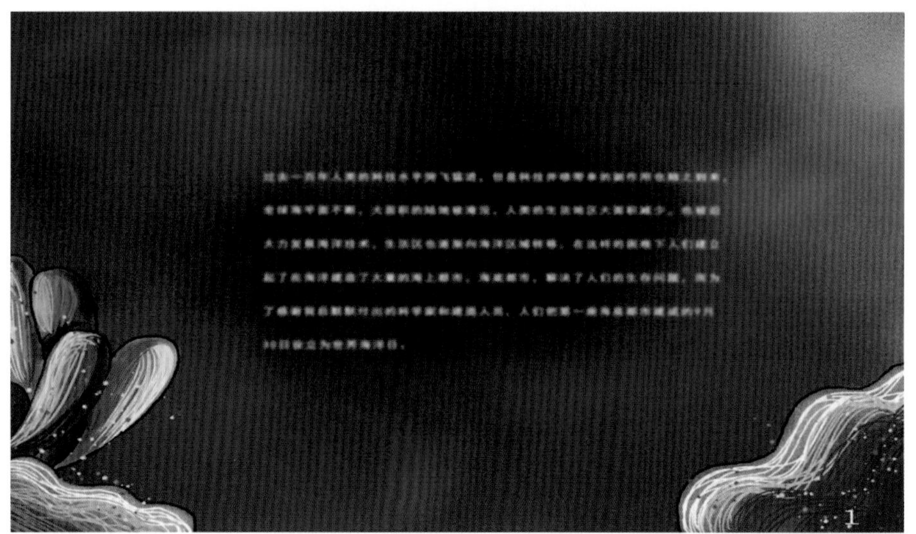

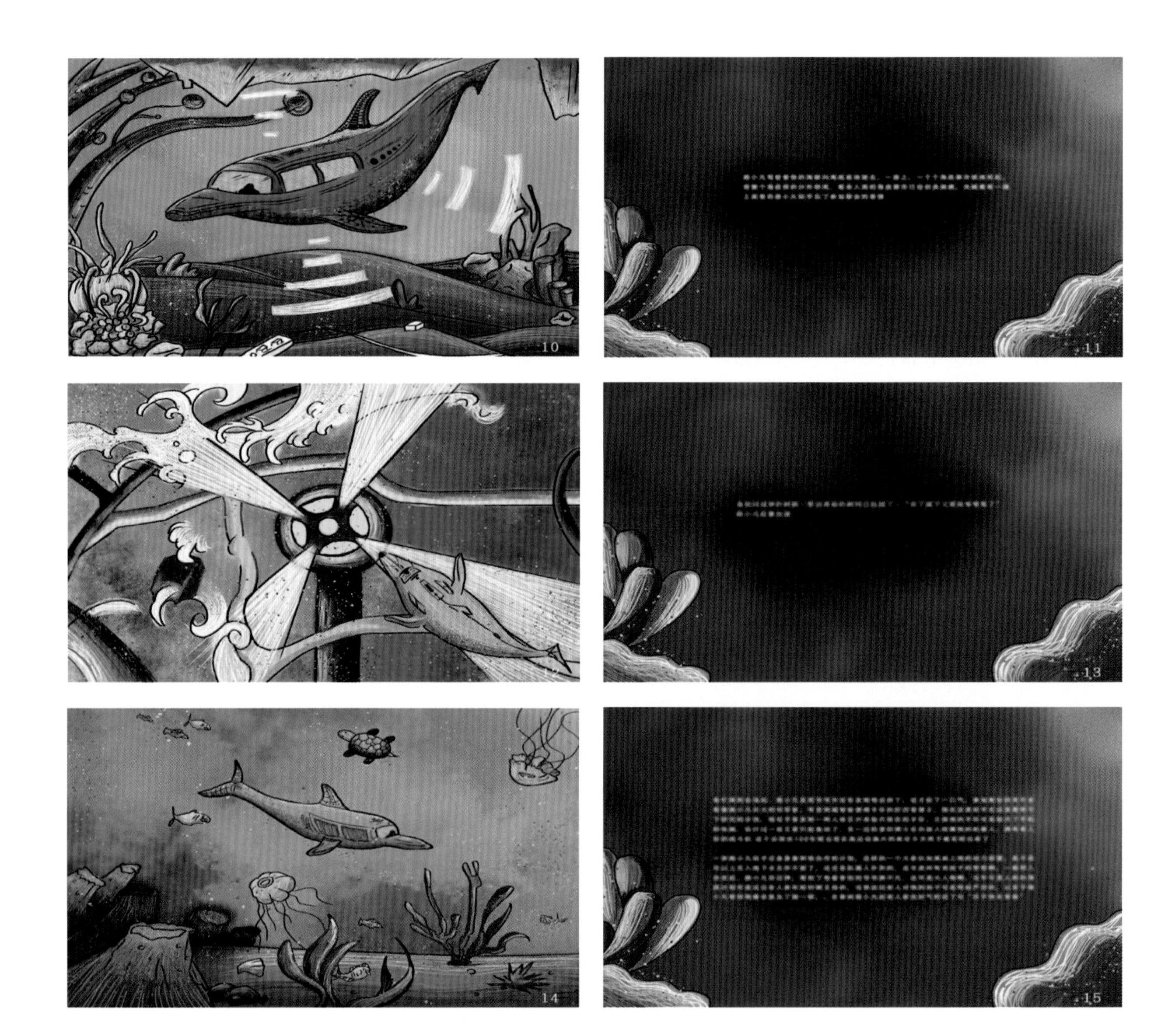


**专家点评：**

《梦想巡航》中，主人公路小凡是一个年轻的天才科学家，在海底城市奠基人、爷爷路不凡以及好厉害博士等科学家的帮助下，他研发的“华夏”号战舰为人类与海洋的未来做出了巨大贡献。绘本为路小凡和其他科学家创设了一场华夏国滴海海底基地的聚会，通过故事化的场景描述、鲜明生动的绘画风格，向读者展现了为海洋生态研究和生态文明建设做出巨大贡献的科学家的风采，激励读者学习他们潜心科研的精神。

## Expert Comments:

The protagonist in the *Dream Cruise*, Lu Xiaofan, is a young genius scientist. With the help of the undersea city founder, grandpa Lu Bufan and Doctor Hao Lihai, etc., his development of the Huaxia warship has made a great contribution to the future of mankind and the ocean. The picture book created a gathering of Lu Xiaofan and other scientists for the Hua Xia ocean seabed bases. By the story-like scenes, vivid painting style and mature painting techniques, it shows readers the elegant demeanor of scientists who have made great contributions to marine ecological research and ecological civilization. It inspires readers to learn from their dedicated research spirit and perseverance.

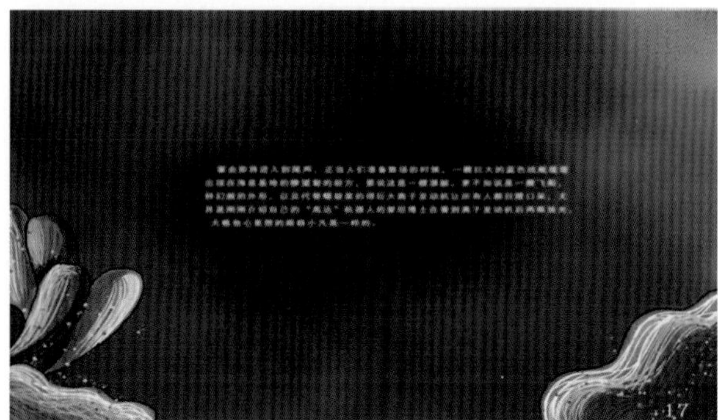

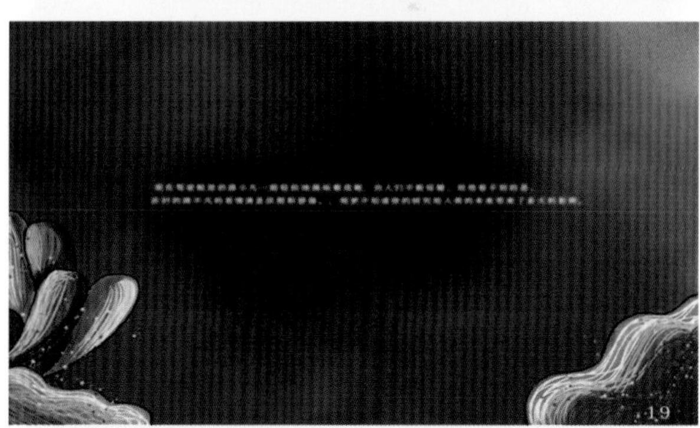

《摩西教授的海洋之约》青岛金门路小学

*Professor Munsey's Ocean Covenant*
Qingdao Jinmen Road Primary School

蓝鲸战队
赵德铭　刘栩菲　曲梓萌

Blue Whale Team
Zhao Deming, Liu Xufei, Qu Zimeng

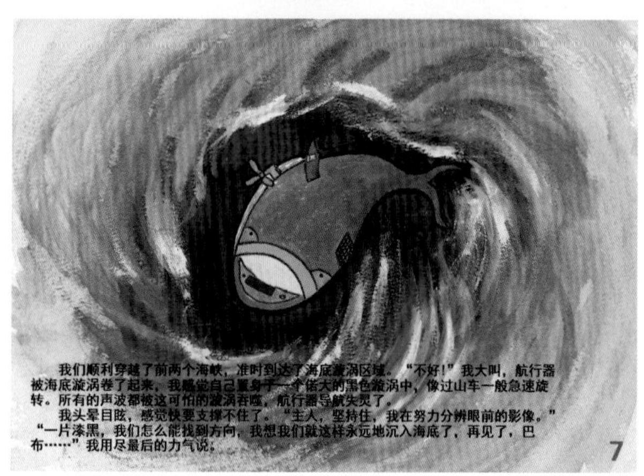


**专家点评：**

《摩西教授的海洋之约》畅想了未来的海洋世界。主人公对摩西教授失踪的那片海域充满了好奇，其驾驶海洋航行器——仿生蓝鲸开启了奇妙之旅。绘本通过对经纬度识别仪、无线传感器等海洋工程装备的描绘将相关海洋科学知识逐一呈现。绘本色彩丰富，文字简洁易懂，为读者插上了想象的翅膀，可激发读者探索海洋世界的兴趣。

## Expert Comments:

*Professor Munsey's Ocean Covenant* imagined the future of the sea world, which is full of curiosity about the waters where Professor Munsey had disappeared. The young author uses the first person tone to drive the ocean vehicle of the simulated octopus Babu to start a wonderful journey. The picture book uses bionic design, longitude and latitude identification device, wireless sensor and other ocean engineering equipment to reflect the knowledge of ocean science in the picture book. The picture book is rich in colors and simple in words, which give readers wings of scientific imagination and help children to explore the mysterious ocean world.

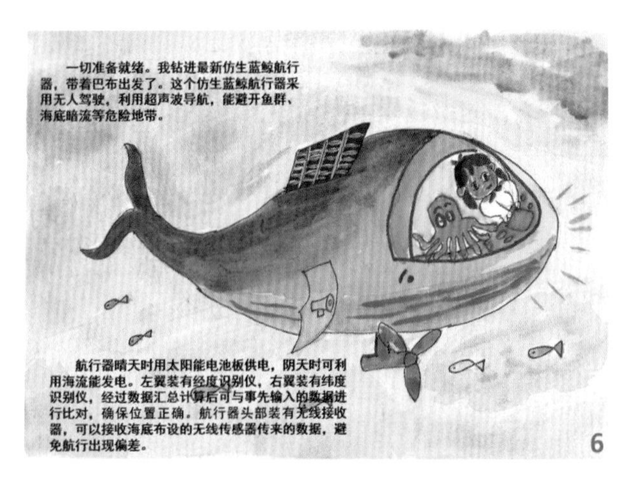

## 《梦殇——来自深海的漂流瓶》中国海洋大学

*Meng Shang — A Drifting Bottle From the Deep Sea*
Ocean University of China

四仙过海
于书萌　张尧榕　许洛川　黄志昊
Four Immortals Cross the Sea
Yu Shumeng, Zhang Yaorong, Xu Luochuan, Huang Zhihao

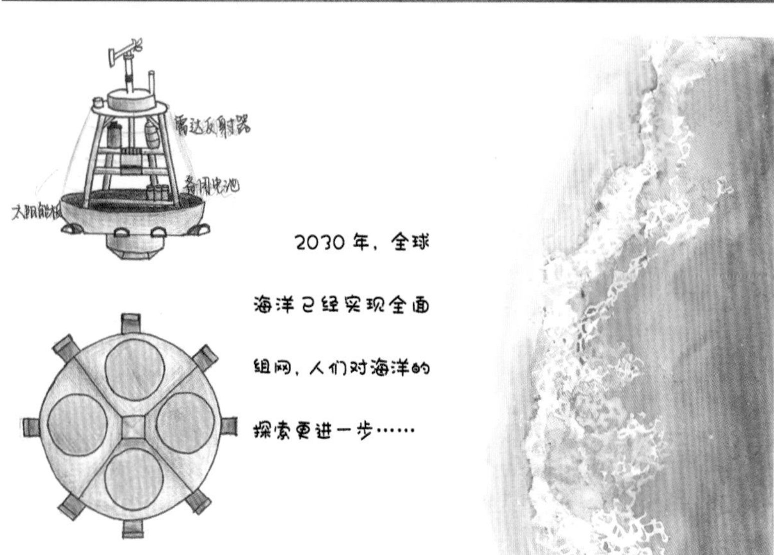

**专家点评：**

《梦殇——来自深海的漂流瓶》讲述了黄萌、张川与青岛分队在随“东方红 2”号科考过程中偶然发现了一个漂流瓶的奇幻故事。他们别出心裁，利用鳗鱼心跳缓慢而精准的特点计时，探索深海的奥秘，展现了科学家不懈探索的精神。“大洋深浅不怕难，敢取透明小深蓝”，绘本绘画风格清新活跃，在体现科技创新成果的基础上展现了文艺风格。

2030 年 6 月 30 日，“东方红 2”号劈波斩浪前往马尼拉海沟，黄萌和张川明日将在马尼拉海沟与青岛分队汇合，为此他们准备了一夜。

科考船捞到一个漂流瓶。

在去往海沟的路上，一条鳗鱼跃上了甲板。深海使它的心跳变得缓慢而精准，可把它放在耐高压的透明圆盘中做成鳗鱼钟。“东方红 2”号依靠深海北斗系统和黑烟囱航标导航。

青岛分队接到深潜小队的汇报后，即刻从青岛站出发，并携带最先进的仪器装备到马尼拉海沟与他们汇合。

他们乘坐的由柔性材料和液态电路制成的航行器，可根据海流情况改变外观，以最大限度利用海流的能量。其顺流时为流线型，逆流时变为特型，使海流对航行器的有用功多于无用功。在潜入深海时，其变为球形，以最大限度承受海底压力。

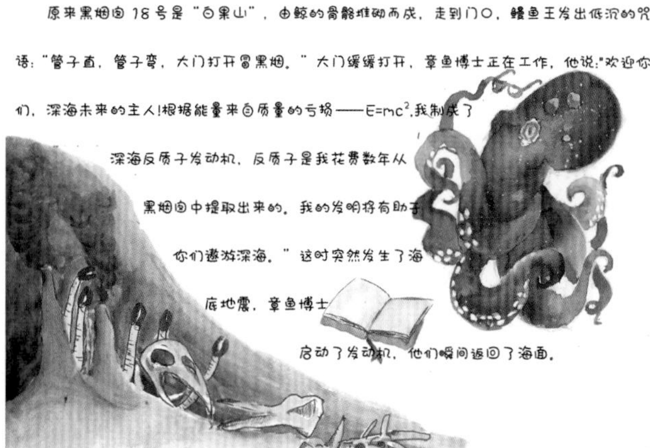


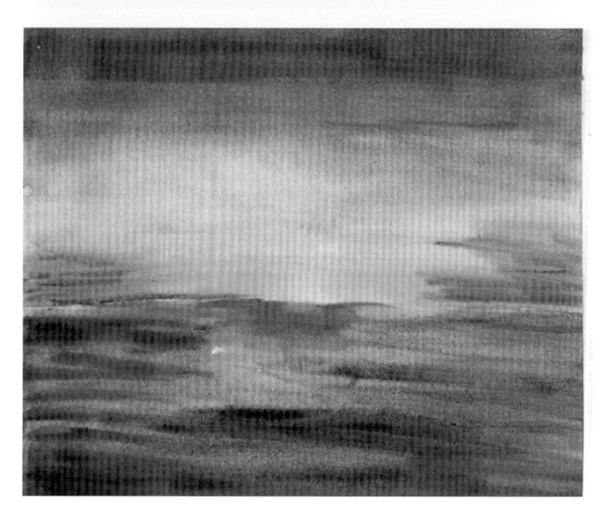

阳光唤醒了张川，他身边的黄萌还在喃喃地说道：“章鱼博士，章鱼博士！”原来他们做了同一个梦！

太阳跃出了地平线，照亮了万顷碧涛，这正是：大洋深浅不怕难，敢取透明小深蓝！

## Expert Comments:

*Meng Shang —A Drifting Bottle From The Deep Sea* tells a fantastic story about Huang Meng, Zhang Chuan and the Qingdao Team accidentally finding a drifting bottle in the process of an expedition with Dongfanghong No.2. They were creative and took advantage of the slow and precise heartbeat of eels to keep time and explore the secrets of the deep sea, which demonstrated the perseverance of scientists. “It is not afraid of the depth of the ocean, but dare to get a transparent small blue portion of the ocean.” The painting style of the picture book is fresh and active, showing the artistic style on the basis of reflecting the achievements of scientific and technological innovation.

## 《我和 Tau Ceti F 星球人的深海约会》青岛市南区实验小学

*My Deep-Sea Date with Tau Ceti F*
Qingdao Shinan Experimental Primary School

深海幽兰
刘珈聿　陈一航
Deep Sea Blue
Liu Jiayu, Chen Yihang

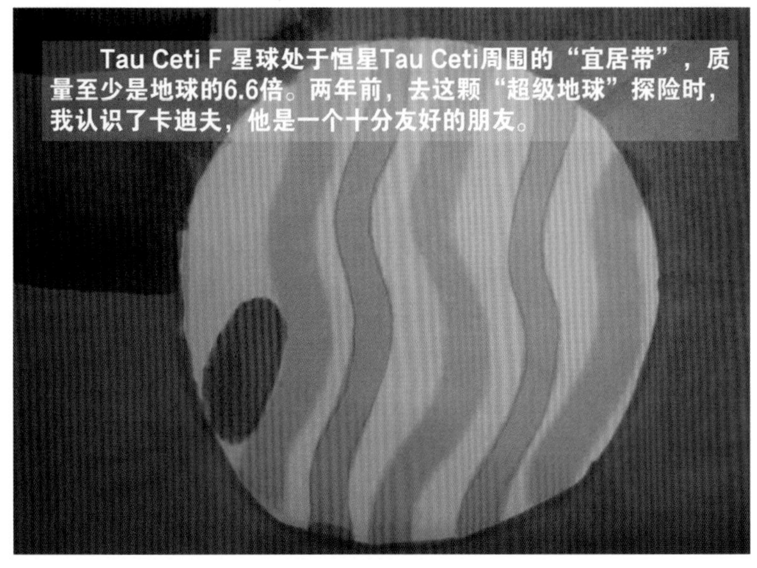

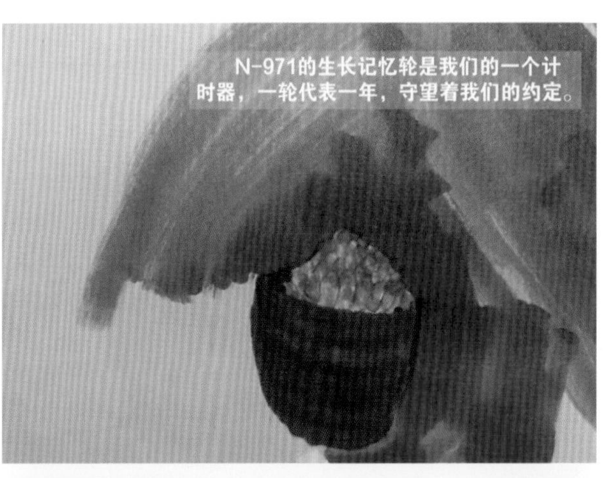


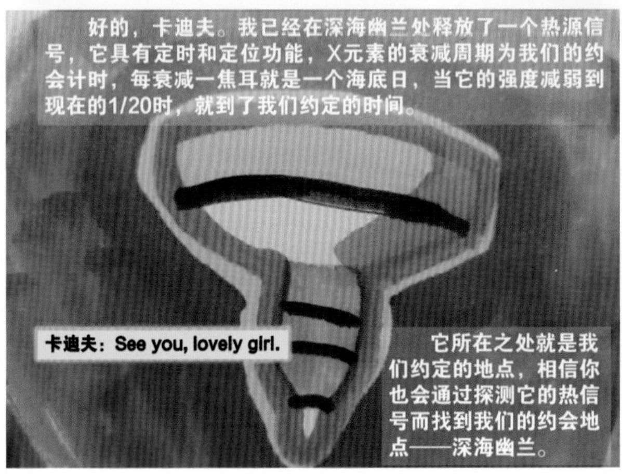


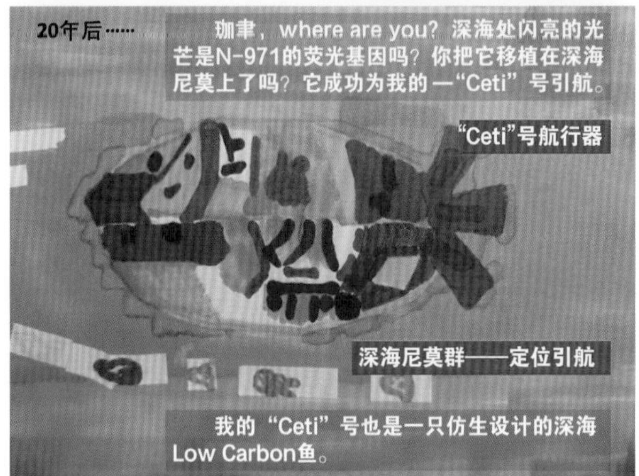


**专家点评：**

《我和 Tau Ceti F 星球人的深海约会》用第一人称的口吻，讲述了一个充满爱和探索精神的小女孩与 Tau Ceti F 星球人卡迪夫在深海幽兰相遇、探索海洋世界的经历。绘本中 N-971 的生长记忆轮象征着爱与希望，用移植了 N-971 的荧光基因的深海尼莫群为航行器引航，体现了“爱为黑暗送去光明”的主题。绘本采用水彩和彩铅进行手绘，以对话的形式，将“科技与爱相融合”寓于其中，丰富了每一个读者的想象。

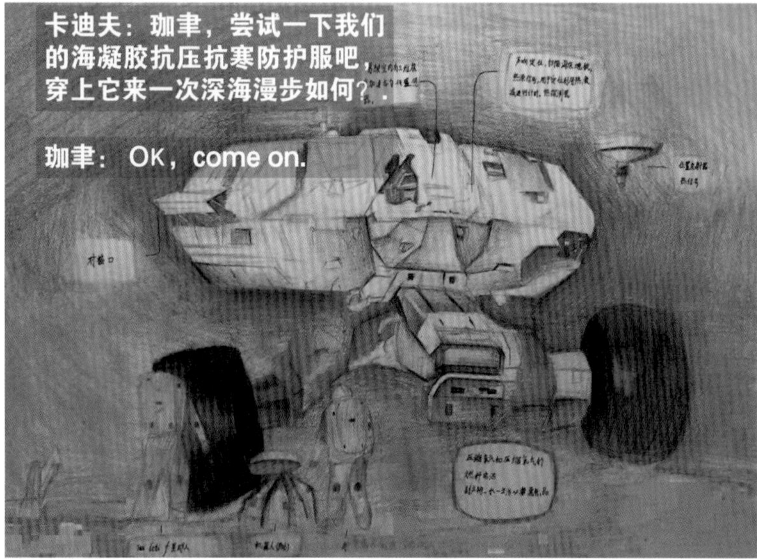


## Expert Comments:

In a first-person tone, *My Deep-Sea Date with Tau Ceti F*, tells the story of a little girl, who is full of love and has an exploration spirit, meeting Tau Ceti F planet man Cardiff in the deep sea to explore the ocean world. In the picture book, the growth memory wheel of N-971 symbolizes love and hope. And the deep-sea Nemo group transplanted with N-971 's fluorescent gene is used as the pilot of the vessel, reflecting the theme that love sends light to darkness. The picture book is hand-painted with watercolor and colored pencils, which integrates science and technology with love in the form of dialogue and enriches the imagination of each reader.

# 设计类竞赛规则及特等奖作品

Competition Rules and Grand Prize Works
in Design Category

### 1.比赛题目

设计一款海洋时空下的智能海洋航行器，其除了具备基本的水中运动、导航定位、环境感知等功能外，还可搭载至少3名乘客，并具有智能化的属性，能更好地满足某项海洋应用的需求。

### 2.比赛形式

海洋工程装备创意设计与演示。

### 3.比赛场地

预赛在网上进行，决赛在中国海洋大学进行。

### 4.比赛说明

1）参赛对象

社会组：已毕业在职人员；

大学组：在校研究生、本科生、专科生；

中小学组：在校高中生、初中生、小学生。

每组队员不超过5人。

2）参赛作品要求

预赛上传设计图（包含三视图、渲染效果图及说明），A3尺寸1页，JPG文件，RGB格式，分辨率300 dpi。决赛打印A3海报1页，如有视频，需为MP4格式，时长不超过2分钟。

3）比赛流程

比赛分为预赛和决赛两个阶段。

预赛阶段，采取专家网评形式，参赛者提交作品电子版，根据评审结果，择优进入决赛。

决赛阶段，入围作品进行路演，每支队伍路演时间不超过5分钟。

### 5.比赛规则

比赛采取评分形式，得分高者获胜。

满分为100分，具体评分标准如下。

（1）要素设计（50分）：海洋航行器应包含动力系统（0分或10分）、导航定位系统（0

分或10分）、环境感知系统（0分或10分）、可搭载至少3名乘客（0分或10分）、智能化功能（0分或10分）。

（2）题目的主旨设计（1~10分）。

（3）航行器结构设计（1~10分）。

（4）航行器材料设计（1~10分）。

（5）航行器工艺设计（1~10分）。

（6）智能化的创意（1~10分）。

对专家的评分进行统计，去掉一个最高分，去掉一个最低分，其余评分取平均值，作为该作品的最终得分。

## 1.Description

Design an intelligent marine vehicle in space-time of ocean, which is available to carry at least three passengers in addition to basic functions including underwater operation, navigation positioning and environmental awareness, and meanwhile it should be intelligent to better meet the needs of a marine application.

## 2.Form

Idea and demonstration of marine engineering equipment.

## 3.Venue

The intermediary contest will be held online and the finals will be held at Ocean University of China.

## 4.More Information

1)Participants

Social group: working individuals;

University group: graduate students, undergraduates, and junior college students;

Primary school and high school group: primary school and high school students.

There are no more than 5 players in each group.

2)Requirements for works

Ocean engineering equipment design (including three views, rendering renderings and instructions), which is one page of A3 paper, JPG file, RGB format and 300 dpi resolution. In the intermediary contest, participants are required to submit one page of poster (with A3 paper). The final video shall be within 2 minutes with a file of MP4.

3)Procedure

The competition is made up of the intermediary contest and finals.

For the intermediary contest, experts will be invited to select online. Participants are required to submit electronic versions of the works, and quality-based works will be netted.

With regarding to finals, the finalists will perform roadshows, and each will be given no more than 5 minutes.

## 5.Rules

Scoring mechanism will be used.

The score is 100 points and scoring rules are as follow:

(1) Element (50 points): the power system (0 or 10 points), navigation and positioning system (0 or 10 points), environment sensing system (0 or 10 points), and at least 3 passengers (0 or 10 points) and intelligent function (0 or 10 points).

(2) The subject of the topic (1 to 10 points).

(3) The structure of the aircraft (1 to 10 points).

(4) The material of the aircraft (1 to 10 points).

(5) The technology of the aircraft (1 to 10 points).

(6) The intelligent idea (1 to 10 points).

The highest score and the lowest score will be removed, then the total score of the referee and the average value will be counted as the score of the work.

# 海底常驻型载人潜水器

## ——应用于海底输油管道维护

参赛单位：北京师范大学南山附属学校
队长：闫嵩嵩
队员：胡久　熊鑫　江中桥　肖文谦

本设计的目的是利用常驻型载人潜水器来替代海洋石油支持船，提高海底输油管道维护工程质量，降低维护成本。

### 模式一：载人模式

当潜水器在有人状态下与海底接驳盒连接时，即为载人模式。当有缆ROV确定输油管道出现破损时，可启动载人模式。潜水器从海底上浮搭载驾驶员与工程师后，移至管道破损处，在工程师的控制下完成对管道的维修。

### 模式二：有缆ROV模式

当潜水器在无人状态下与海底接驳盒连接时，即为有缆ROV模式。无缆AUV在日常巡检中发现输油管道存在可疑破损点时，可启动有缆ROV模式。工程师在控制中心远程实时控制潜水器，对输油管道的可疑破损点进行全方位的观察与判别。

### 模式三：无缆AUV模式

当潜水器没有与海底接驳盒连接时，即为无缆AUV模式。无缆AUV主要是对海底输油管线进行日常巡检。巡检完成后，其自动将巡检数据通过接驳盒传至岸上控制中心。工程师根据日常巡检数据，分析输油管道可能存在的问题。无缆AUV模式采用短基线法在海底进行导航与定位。

海底接驳站

油气管道

## 《海底常驻型载人潜水器》北京师范大学南山附属学校

*Permanent Submarine Manned Submersible*
Nanshan Affiliated School of Beijing Normal University

**新栈道队**
**闫嵩嵩　胡久　熊鑫　江中桥　肖文谦**
The New Road Team
Yan Songsong, Hu Jiu, Xiong Xin, Jiang Zhongqiao, Xiao Wenqian

**专家点评：**

海底常驻型载人潜水器设计，针对海底输油管道维护的传统方式提出了独特的设计解决方案，具有领先性特征，主旨创意明确，具有较高的创新性。该潜水器外观设计采用仿生流线造型，体现了时代感；技术方案可以实现，智能化水平较高；主体结构较为合理，选用材料考究；但在设计表现力方面还需进一步提升，同时也要考虑潜水器自身的维护。

**Expert Comments:**

The design of permanent submarine manned submersible puts forward its own unique design and innovation solution for the problem of the traditional ocean maintenance way of the submarine oil pipeline, which has the leading characteristics, clear theme of creativity and high innovation level.

The appearance design adopts bionic streamline modeling, reflecting the sense of the times. The technical scheme can be realized, and the intelligence level is high. The main structure is reasonable, and the selection of materials is exquisite. However, the design performance needs to be further improved, and also needs to consider the maintenance of the submersible itself.

主视图　　左视图　　俯视图

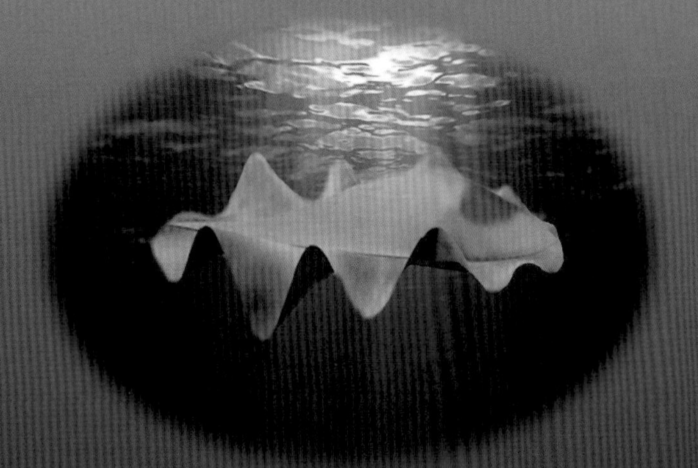

海底效果图

陆地效果图

# "鲲鹏"号 仿生波动鳍推进的两栖航行器

载人耐压舱

蓄电池箱

合金壁

观测窗

推进器

摄影摄像机

测距侧扫声呐

探照灯

## 《“鲲鹏”号——仿生波动鳍推进的两栖航行器》重庆交通大学

*“Kun Peng” Model — Bionic Undulating Fin Propulsion Amphibious Vehicle*
Chongqing Jiaotong University

**重点团队**
**谢萍　牛浩南　黄寅杰　马灿　李文浩**
Key Team
Xie Ping, Niu Haonan, Huang Yinjie, Ma Can, Li Wenhao

### 专家点评：

《“鲲鹏”号——仿生波动鳍推进的两栖航行器》是根据仿生学原理设计的创新方案，旨在解决海洋环境和海洋滩涂环境的治理问题，有一定的创新性和现实意义。采用的技术方案基本可行，智能化水平有所体现；结构设计在水中比较合理，但对在滩涂中的应用没有深入考虑；根据材料选择仿生造型，体现出设计时尚性特征，但材料的可实现性需进一步研究；设计表现力和美感需进一步优化。

### Expert Comments:

*“Kun Peng” Model — Bionic Undulating Fin Propulsion Amphibious Vehicle* is an innovative design scheme based on bionics principle, aiming at solving the governance of marine environment and marine tidal flat environment. It has a certain degree of innovation and practical significance. The technical scheme adopted is basically feasible and the intelligence level is reflected. The structure design is reasonable in water, but not in tidal flat. Bionic modeling is selected according to the material, which reflects the fashion characteristics of the design, but the realization of the material should be further studied. The design expressiveness and aesthetics need to be further optimized.

# “THUNDER”助力者号

## 智能水下机器人投放装置

## 三视图

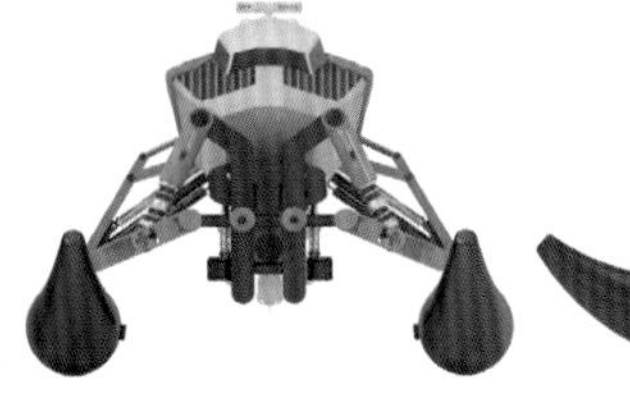

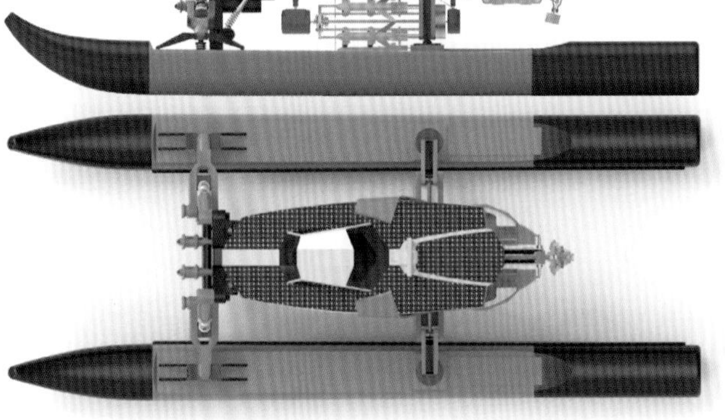

THUNDER

- **船长：** 23.7米
- **船宽：** 9.4米
- **船高：** 7.1米
- **载重量：** 48.1吨

## 设计过程

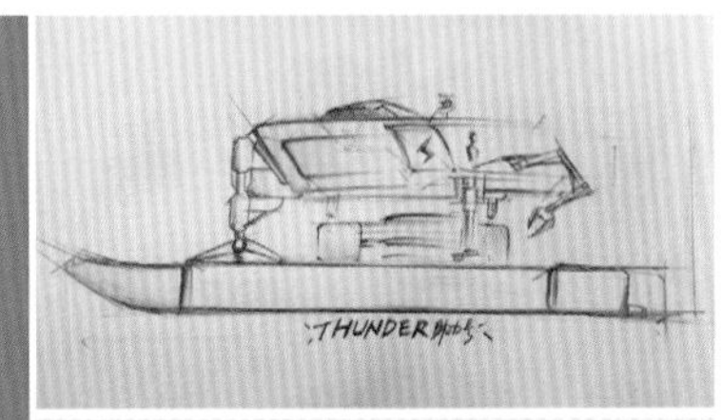


草图构思

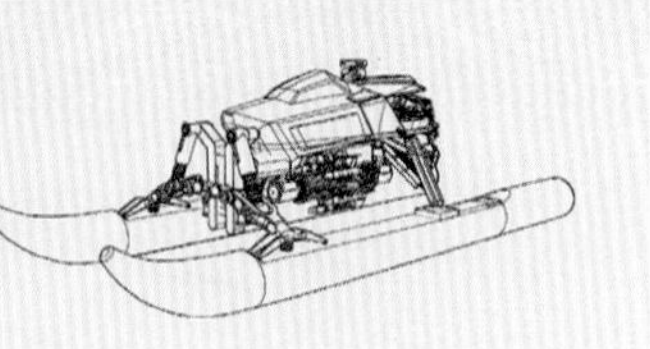

建模线稿

渲染效果

## 简介

**一款经济实用、投放回收效率高的智能水下机器人投放装置**

旨在弥补水下滑翔机现有投放回收方式的不足；
为水下滑翔机的投放回收提供新思路和新方法；
未来依靠补给船的配合可实现多个水下机器人投放装置的编队作业。

## 创新设计

**智能化的水下机器人投放和回收**

采用滚轮式投放、机械臂回收、履带传输置位、液压升降维修的设计原理，实现机器人的投放、回收、维修和二次投放。

**环境感知系统**

基于APM的智能航行，具有航行、作业、停泊等功能，可同时实现人工和自动操作干预模式。

**抗风浪结构创新设计**

双体船平台与浮筒采用铰接支架和弹簧连接的方式，降低投放设备意外事故出现的概率，增强双体船抗风浪的能力。

**动力系统**

电力驱动双喷水推进使双体船具有卓越的高速机动性，改善了其水动力特性，提高了空间利用率和推进器的使用效率。

**导航定位系统**

基于北斗导航系统的覆盖范围和信号频率，利用机械手上的可视化摄像头，提供精确定位和高效回收作业。

**可搭载至少3名乘客**

清晰的驾驶和操作区域，合理的空间布局，可有效提高作业效率（同时可实现无人化操作）。

广东海洋大学

THUNDER

## 《“THUNDER”助力者号智能水下机器人投放装置》广东海洋大学

*"THUNDER" the Helper Model of Intelligent Underwater Vehicle Delivery Device*
Guangdong Ocean University

**广东海洋大学一队**
**胡琼蕾　宋子洋　陈沛楷　张健超**
1st Team of Guangdong Ocean University
Hu Qionglei, Song Ziyang, Chen Peikai, Zhang Jianchao

**专家点评：**

“THUNDER”助力者号智能水下机器人投放装置整体设计完整，思路清晰，船体结构合理，系统较完整，技术路线明确，可实现转化，应用范围可拓展。造型美观新颖，充分考虑了工艺布局的合理性，设计极具时代感和竞争力。

### Expert Comments:

The overall design of “THUNDER” the Helper Model of Intelligent underwater vehicle delivery device is complete. The innovative design is clear, the hull structure is reasonable, and the system is relatively complete. The technical route is clear and can be transformed, the scope of application can be expanded. The appearance is pretty and novel, and fully considered the reasonable layout of the process. The design performance is very modern and competitive.

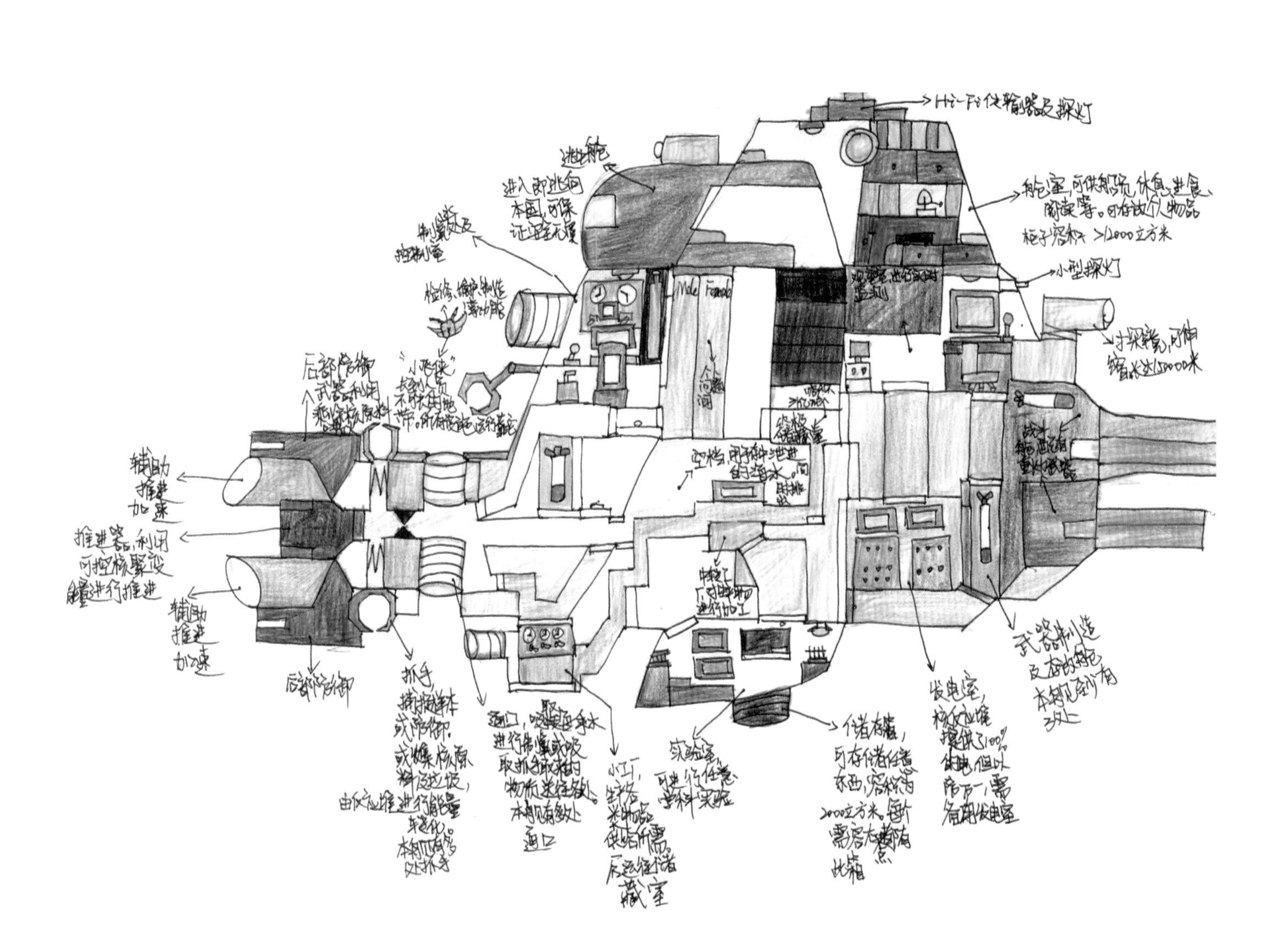
Hi-Fi传输器及探灯
小型探灯
“小飞侠”
检修、维护、制造等功能
辅助推进加速
推进器，利用可控核聚变能量进行推进
辅助推进加速
后部防御
抓手
实验室，可进行任意学科实验
发电室，核反应堆提供了100%的电，但以防万一，需备用发电室

**《舰艇》青岛银海学校**

*Warship*
Qingdao Yinhai School

**精灵潜艇**
**刘孝衍**
The Elves Submarine
Liu Xiaoyan

**专家点评：**

舰艇创新体现出较理想的概念设计，充分展示出设计者的想象力和创新能力，同时展现出较强的设计表达能力。整体采用仿生原理进行设计，技术路线较为明确，舰艇结构较为合理；选材方面设计者没有考虑。设计者若能用计算机辅助设计进行表现，设计创新效果会更加优秀。

Expert Comments:

The ship's innovation embodies the ideal concept design, gives full play to the designer's imagination and innovation ability, and shows a strong ability of design expression. The whole model adopts the bionic principle to design. The technical route is clear and the structure of the ship is reasonable. However, material selection was not considered. If the designer could use computer aided design to perform, the design innovation effect would be more excellent.

# 制作类竞赛规则及特等奖作品

Competition Rules and Grand Prize Works
in Production Category

**1.比赛题目**

制作高速海洋水下潜航器模型。该模型在水中接近中性（允许±10%的偏差），具有水下推进的功能，应用了海洋潜航器必备防腐、耐压、水密等技术。水动力和布局设计合理，具有较低的水阻，且做工精良。

**2.比赛形式**

图文资料展示、现场水中演示、测试与现场展示相结合。

**3.比赛场地**

预赛在网上进行，决赛在中国海洋大学进行。

**4.比赛说明**

1）参赛对象

社会组：已毕业在职人员；

大学组：在校研究生、本科生、专科生；

中小学组：在校高中生、初中生、小学生。

每组队员不超过5人。

2）参赛作品要求

网评提供设计制作说明书和视频，说明书为A4尺寸的PDF文件，大小不超过5 M；视频格式为MP4，时长不超过1分钟，文件大小不超过10 M。

决赛提交作品模型、设计制作说明书。实物模型质量小于20 kg（辅助结构质量除外），模型长、宽、高之和小于158 cm，模型说明书不超过20页（A4尺寸）。

3）比赛流程

比赛分为预赛和决赛两个阶段。

预赛阶段，采取专家网评形式，参赛者提交作品模型照片、设计说明书和视频，作品择优进入决赛。

决赛阶段，参赛者提交作品模型、设计说明书和决赛视频，进行现场路演，每支队伍路演时间不超过5分钟。

### 5.比赛规则

比赛采取评分形式，得分高者获胜。

满分为100分，具体评分标准如下。

（1）浮力（10分）：水中重量为空气中重量±10%（0分或10分）；

（2）防腐（10分）：至少具备一种防腐措施（0分或10分）；

（3）水密（10分）：至少具备一种密封方式（0分或10分）；

（4）耐压（10分）：至少具备一种耐压增强措施（0分或10分）；

（5）运动（10分）：具备水下可动的推进装置（0分或10分）；

（6）减阻（15分）：每项减阻措施5分，最高15分；

（7）整体布局（1~10分）；

（8）水动力设计（1~10分）；

（9）做工精细程度（1~10分）；

（10）美观程度（1~5分）。

对专家的评分进行统计，去掉一个最高分，去掉一个最低分，其余评分取平均值，作为该作品的最终得分。

## 1.Description

To produce a high-speed ocean underwater submersible model. The model, close to neutral in the water (±10% deviation is permitted), plays a role in propulsion underwater, and it also has advantages of anti-corrosion, pressure and water tightness for marine submersibles. The design of hydrodynamic and layout is reasonable with low water resistance and excellent technology.

## 2.Form

Graphic description, demonstration and testing in the water and live show.

## 3.Venue

The intermediary contest will be held online and the finals will be held at Ocean University of China.

## 4.More Information

1)Participants

Social group: working individuals;

University group: graduate students, undergraduates, and junior college students;

Primary school and high school group: primary school and high school students.

There are no more than 5 players in each group.

2)Requirements for works

The online review instruction should be within 5M with A4 paper, PDF file.The online review video should be within 1 minute with size of less than 20M, MP4 file.

The group must submit the work model and design and production manual in the finals. The model should be less than 20 kg (excluding the auxiliary mechanism) with the sum of the length, width and height of less than 158 cm; the manual should be finished within 20 pages (with A4 paper).

3)Procedure

The competition is made up of the intermediary contest and finals.

For the intermediary contest, experts will be invited to select online. Participants are required to submit model photos of works, design brochures and online review videos and quality-based works will be netted.

With regarding to finals, the finalists will perform roadshows about their model photos of works, design brochures and online review videos, and each will be given no more than 5 minutes.

## 5.Rules

Scoring mechanism will be used.

The score is 100 points and scoring rules are as follow:

(1) Buoyancy (10 points): The weight in water accounts for ±10% of air (0 or 10 points);

(2) Anti-corrosion (10 points): at least one way of anti-corrosion (0 or 10 points);

(3) Water tightness (10 points): at least one way of sealing (0 or 10 points);

(4) Anti-pressure (10 points): at least one way of anti-pressure (0 or 10 points);

(5) Movement (10 points): underwater propulsion device is needed (0 or 10 points);

(6) Resistance reduction (5 points): each scores 5 points with the cape of 15 points;

(7) Overall layout (1 to 10 points);

(8) Hydrodynamic design (1 to 10 points);

(9) Accuracy of technology (1 to 10 points);

(10) Appearance (1 to 5 points).

The highest score and the lowest score will be removed, then the total score of the referee and the average value will be counted as the score of the work.

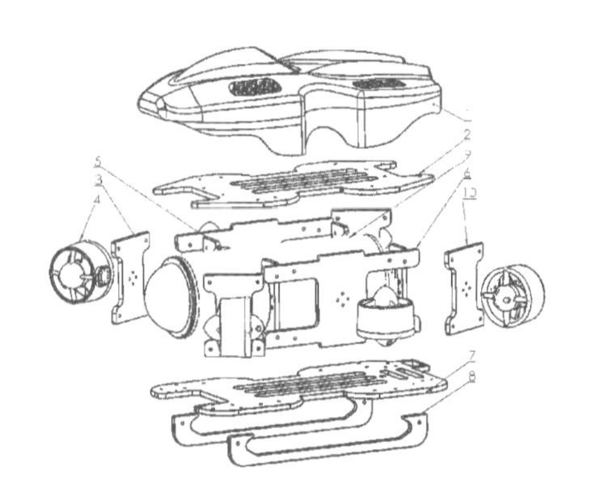

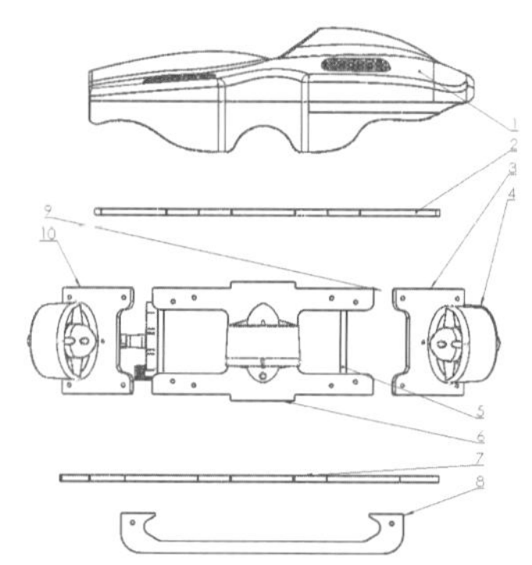

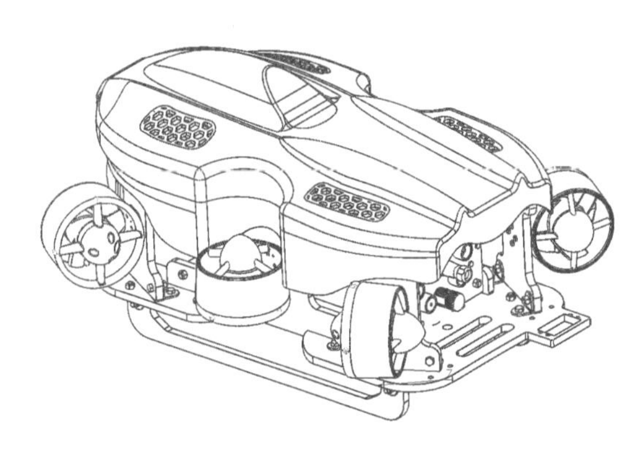

## 《“哪吒”号智能水下巡检机器人》郑州大学

*“Nezha” Intelligent Underwater Inspection Robot*
Zhengzhou University

**“哪吒”传奇**
**卢航宇　刘德威　明瑞烨　段嘉璐　谷炎达**
The Legend of “Nezha”
Lu Hangyu, Liu Dewei, Ming Ruiye, Duan Jialu, Gu Yanda

**专家点评：**

此作品针对水下巡检任务，对机器人本体结构进行了科学的计算和分析，实现了航行器“零浮力”和“零重力”的特性，并且定量计算和设计了机器人浮心和重心的距离以提高其航行的稳定性，机身做工精美，在水密、防腐、减阻等方面也做出了高质量的设计。综上，本作品体现了参赛选手扎实的工程技能和优良的科学素养。

Expert Comments:

Aiming at the underwater inspection task, this work has carried out scientific calculation and analysis on the body structure design of the robot. It realized the characteristics of zero buoyancy and zero gravity of the vehicle, and quantitatively calculated and designed the distance between the center of buoyancy and the center of gravity of the robot to improve the stability of its navigation. The fuselage is not only exquisite, but also presents high quality in water tight, anti corrosion, drag reduction and some other aspects. In conclusion, this work reflects the solid engineering skills and excellent scientific literacy of the contestants.

# 工程类竞赛规则及特等奖作品

Competition Rules and Grand Prize Works
in Engineering Category

## 1. 比赛题目

海底管道巡检：海底管道形状如图1所示，共有9段，共有11个可供选择的腐蚀或漏油点（11个霍尔传感器感应点），各段管道最长为5米，最宽为2.5米（以实际尺寸为准），水池为蓝色，管道为白色，管道搭建在长7.5米、宽3.5米的水池底部。选手的水下航行器底部安装强磁（磁铁采用统一规格N35 20毫米×10毫米×4毫米，可安装2片），航行器从出发区出发，触发A点开始计时，若在出发1分钟之后仍无法触发A点，巡管仍然开始计时。计时后，比赛时间为5分钟，航行器依次经过A点和后面的几个点，岸上有11个灯对应水下11个漏油点，航行器每经过1个点岸上对应的灯便会点亮，5分钟到后，灯不会再被点亮，比赛结束；或11个漏油点在5分钟之内被全部找到，计时结束。

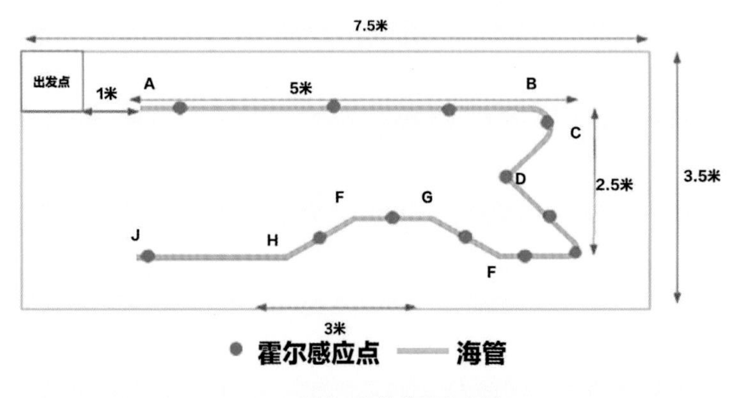


图1 场地布置参考图

## 2. 比赛形式

参赛者自行制作、组装或购买水下航行器参赛，设备在空气中的重量不超过35 kg，最大功率不能超过3000瓦，在规定的时间内分数最高者获胜。

## 3. 比赛场地

中国海洋大学，场地布置如图1所示。场地内尺寸：7.5米×3.5米；水深0.8~1米。

## 4. 比赛说明

1）参赛对象

社会组：已毕业在职人员；

大学组：在校研究生、本科生、专科生；

中小学组：在校高中生、初中生、小学生。

每组队员不超过5人。

2）比赛流程

参赛队伍根据任务要求依次在比赛场地完成任务。

**5.比赛规则**

采用遥控方式，每找到1个漏油点即点亮1盏灯，得10分；采用自主方式（如通过蓝色背景和白色管道）每找到1个漏油点即点亮1盏灯，得20分。得分高者获胜，得分相同用时最短者获胜。

## 1.Description

Submarine pipeline inspection: There are 11 oil leakage inspections (Hall sensors) on the submarine pipeline (Figure 1), which is made up by 9 sections. The maximum length and width of the pipeline are 5 metres long and 2.5 metres wide (based on actual size). The pool is blue background and the pipeline is white. The pipeline is built at the bottom of the pool, which is 7.5 metres long and 3.5 metres wide. The bottom of the underwater vehicle operated by the player is equipped with a strong magnet (with twos of uniform specifications N35 of 20 mm×10 mm×4 mm). The underwater vehicle departs from the starting area and once the point A is triggered, the timing starts. Within 1 minute, if the point A is not triggered, the timing still starts. The underwater robot should pass through all the points. In the process, there may be submarine air bubbles and currents. Each light on the shore corresponds to a leakage points in the water and the light will be lit the moment the underwater vehicle passes. Over 5 minutes, the light will not be lit again. If all of the oil leakage inspections are found, the timing is over.

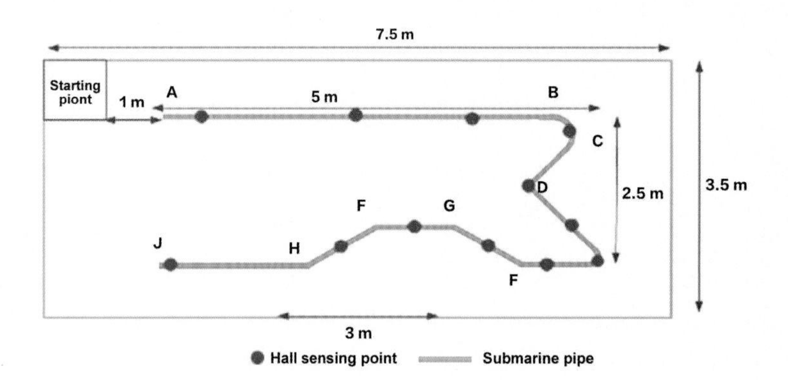


Figure 1　Layout of the venue

## 2.Form

Participants use their underwater robot that can be made by themselves, assembled and purchased to finish the contest. The robot is not allowed to exceed 35 kg with the maximum power of less than 3,000w. Those who gets the highest score in the given time will win.

## 3.Venue

Ocean University of China, the layout is shown in Figure 1. The size of the venue: 7.5 m × 3.5 m; depth of water: 0.8 to 1 meter.

## 4.More Information

1) Participants

Social group: working individuals;

University group: graduate students, undergraduates, and junior college students;

Primary school and high school group: primary school and high school students.

There are no more than 5 players in each group.

2) Procedure

The group should complete the tasks in the venue according to the requirements.

## 5.Rules

Those will score 10 points, if they look for a leakage and a light is lit by a remote control method (for example, looking for pipes through blue background and white pipes). Those will score 20 points, if they look for a leakage and a light is lit by an autonomous way. The group with the highest score wins; if the score is the same, time will matter.

## 《ROV 水下智能机器人》南通理工学院

*ROV Underwater Intelligent Robot*
Nantong Institute of Technology

拓创小队
张艺沥　李爽　陈烨辉　李龙　张梓淇
Tuo Chuang Team
Zhang Yili, Li Shuang, Chen Yehui, Li Long, Zhang Ziqi

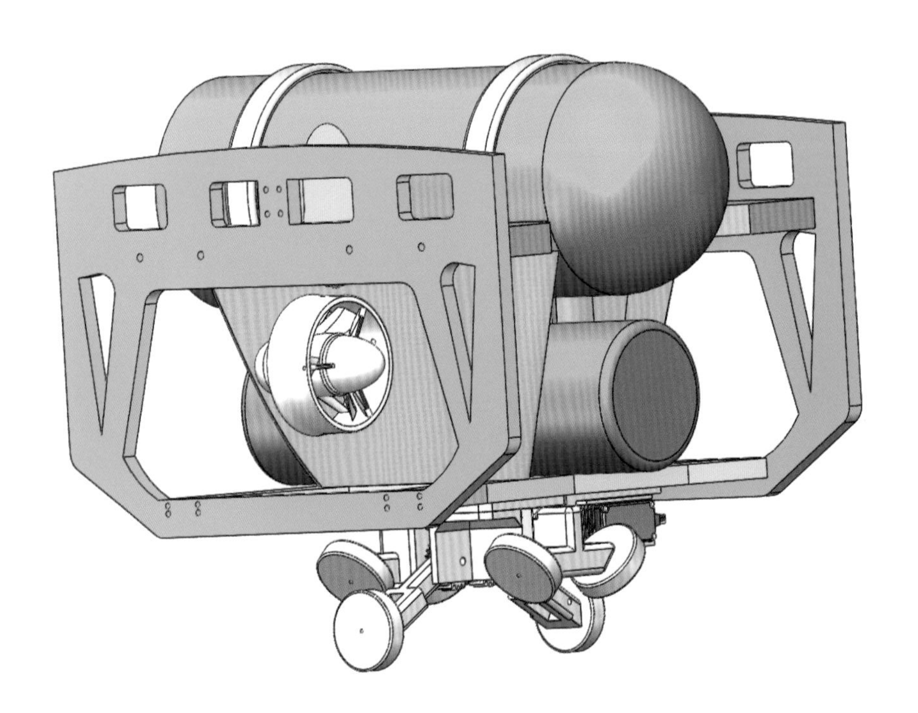

**专家点评：**

南通理工学院代表队的这件作品体现了较强的创新实践能力。这件作品的惊艳表现可以用轻盈灵动来形容。ROV 尺寸不大，在水中的前进速度却不低。队员们配合熟练，小 ROV 就像蜜蜂那样在舞动中前行，调头、转弯都无比灵活。

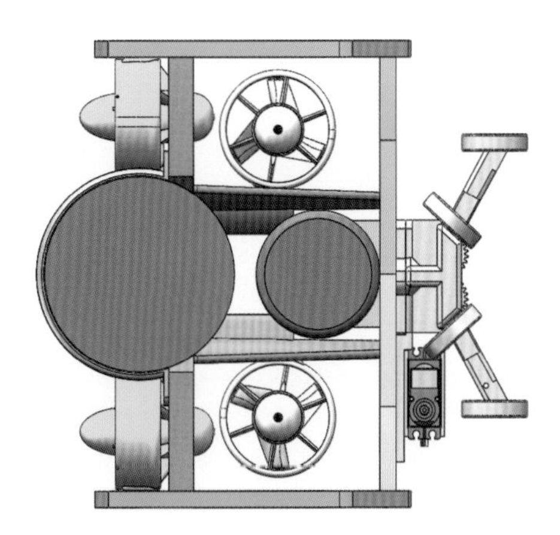

## Expert Comments:

The team from Nantong Institute of Technology has outstanding innovative and practical ability. The stunning performance of this work can be described as light and flexible. The size of the ROV is not big, but the forward speed in the water is not low. The team members cooperated well, and the ROV moved forward in dancing just like a bee—turning around with great agility.

**《极光》海军潜艇学院**

*Aurora*
PLA Navy Submarine Academy

**极光**
**刘佳仕　高君建　朱豪泽　王晟洵　叶嘉诚**
Aurora
Liu Jiashi, Gao Junjian, Zhu Haoze, Wang Shengxun, Ye Jiacheng

**专家点评：**

海军潜艇学院的这件作品，方方正正，行动稳重，有一种“军工产品”的可信赖感。比赛时，队员们根据赛场情况对探测器的外罩进行了改造，使之在外形上能够更加贴近被测管线。这一做法显然收到了很好的效果。水池中的ROV稳扎稳打，一步一个脚印，沿着管线稳步前进，感应点就一个个出现在监测屏幕上了。流畅的感觉就连专家与观众也觉得十分惬意。

Expert Comments:

The work from the PLA Navy Submarine Academy is square in shape and steady in action, bringing a sense of trust in military products. During the competition, the team modified the outer cover of the probe to make it more closely resemble the testing pipeline. This approach has apparently achieved good results. The ROV in the pool moves steadily, step by step, moving along the pipeline, and the sensor points appear on the monitoring screen one by one. The smooth feeling makes even the referees and the audiences feel very pleasant.

第二届国际海洋
工程装备科技创新大赛
The 2nd International Marine Engineering
Equipment Science and Technology
Innovation Competition

**大赛主题：海洋跨介质航行**

海天相映亘古不变，海洋奔腾不息。虽然海洋与天空都同样广阔，但是由于海水与空气的介质特性不同，所以对航行器产生了不同的流体力学作用。如今随着科技的发展与创新能力的提升，人类向海图强的愿望日益强烈。海陆隔阂已被打开，海天跨界等待着我们去畅想。因此本届大赛以海洋跨介质航行为主题，进行跨介质航行器相关的创意科幻绘本制作与概念设计竞技。

Theme : Ocean Trans-media Sailing

The sea and the sky reflect each other, the sea flows without stopping. Although the sea and the sky have the same expanse, due to the different medium characteristics of sea water and air, there are different hydrodynamic interactions for the vehicle. Nowadays, with the development of science and technology and the improvement of innovation ability, the desire of human beings to make achievements on the ocean is increasingly strong. The barrier between land and sea has been opened, and how to break the boundary between sea and sky is waiting for our imagination. Therefore, this competition will focus on the theme of ocean trans-media vehicle, and set up a creative science fiction picture book and concept design competition related to trans-media vehicle.

# 科幻类竞赛规则及特等奖作品

Competition Rules and Grand Prize Works
in Science Fiction Category

**1.比赛题目**

想象在未来的2070年，组织一次海洋跨介质（海洋、天空，注意不是太空）旅行。设想的航行器要能在海水中、海面、天空航行。构想一种或几种载人航行器方案，使其不仅适用于海洋环境，还适用于天空环境。

构想至少两种装置或技术，使其应用于该航行器，使得这次旅行更加方便。采用绘本形式的“硬科幻”（以表现科学技术为重点的科幻作品），尽量用图来表达，注重体现技术含量。

**2.比赛形式**

科幻绘本。

**3.比赛说明**

1）参赛对象

大学组：在校研究生、本科生、专科生；

中小学组：在校高中生、初中生、小学生；

每组队员不超过4人。

2）参赛作品要求

围绕大赛主题，以海洋跨介质旅行为故事背景绘制作品，形式为手绘或计算机绘图，字数、篇幅不限，上传文件为PDF格式，文件大小不超过15 M。

3）比赛流程

根据国家与学校防疫要求，赛事全程在网上进行，比赛分为预赛和决赛两个阶段。

预赛阶段，采取专家网评形式，参赛者提交作品电子版（手绘作品为扫描版），作品择优进入决赛。

决赛阶段，入围作品进行线上路演，每支队伍路演与专家问答时间不超过10分钟。

**4.比赛规则**

比赛采取评分形式，得分高者获胜。

满分100分，具体评分标准如下。

（1）要素的准确性（30分）：绘本中包含旅行用的航行器（0分或5分），能跨介质的旅行事实（0分或5分），旅行的主角（主角是人，不能采用拟人的方式）（0分或5分），跨介质的原理（0分或5分），易于跨介质旅行的技术1（0分或5分），易于跨介质旅行的技术2（0分或5分）。

（2）作品的创意（50分）：故事情节的整体创意（1~10分）、跨介质原理的科学创意（1~20分）、航行器技术装置1的创意（1~10分）、航行器技术装置2的创意（1~10分）。

（3）作品的文学性（1~10分）。

（4）作品的艺术性（1~10分）。

对专家的评分进行统计，去掉一个最高分，去掉一个最低分，其余评分取平均值，作为该作品的最终得分。

## 1.Description

Imagine organizing a trip across the ocean (ocean, sky: not space) in the future, 2070. The envisioned vehicle should be able to navigate in the sea, on the sea and in the sky. Think of one or more principles that can be adapted not only to the ocean environment but also to the sky environment to construct such a manned vehicle.

Imagine at least two devices or technologies that could be applied to the vehicle to make the trip more convenient. In the form of picture books, “hard science fiction” (science fiction works focusing on science and technology) is expressed as much as possible with pictures, focusing on the technical content.

## 2.Form

To complete a sci-fi picture book that is aimed at the them.

## 3.More Information

1) Participants

University group: graduate students, undergraduates, and junior college students;

Primary school and high school group: primary school and high school students.

There are no more than 4 players in each group.

2) Requirements for works

To complete a science fiction book bases on the story background of ocean trans-media travel. The format is freehand drawing or computer drawing. There is no limit on the number of words or length. The uploaded file should be in PDF format and the size of the file should not exceed 15 M.

3) Procedure

According to the requirements of the state and the school, the whole competition was conducted online, and the competition was divided into two stages: preliminary and final.

In the preliminary stage, online evaluation by experts will be adopted. Participants will submit electronic versions of their works (hand-painted scanned versions), and the best works will enter the final.

In the final stage, the finalists will go through an online roadshow, and each team will have 10 minutes to complete the roadshow and Q&A with experts.

## 4.Rules

Scoring mechanism will be used.

The score is 100 points and scoring rules are as follows.

(1) Elements (30 points): The vehicle for travel in the picture book (0 or 5 points), the fact that it can travel across media (0 or 5 points), the protagonist of the journey (the protagonist is a person, can not be anthropomorphic way) (0 or 5 points), principle of trans-media(0 or 5 points), techniques 1 for easy travel trans-media (0 or 5 points), technique 2 for easy travel trans-media (0 or 5 points).

(2) Creation (50 points): the idea of the story (1 to 10 points), the scientific idea of trans-media principle (1 to 20 points), the idea of technical equipment 1 for vehicle (1 to 10 points) and the idea of technical equipment 2 for vehicle (1 to 10 points).

(3) Literature (1 to 10 points).

(4) Artistic (1 to 10 points).

The highest score and the lowest score will be removed, then the total score of the referee and the average value will be counted as the score of the work.

《小花寻宝记》青岛银海学校

*Little Flower Treasure Hunt*
Qingdao Yinhai School

中小学组
付瑜
Primary School and High School Group
Fu Yu

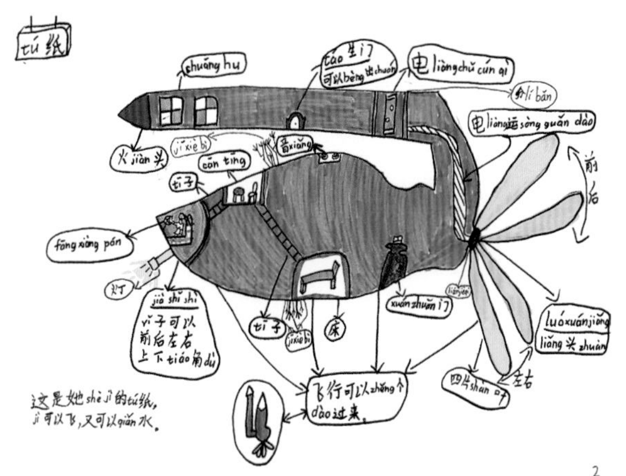

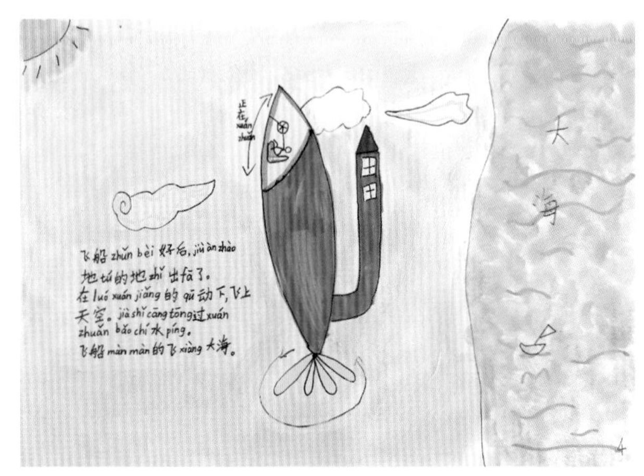

**专家点评：**

《小花寻宝记》故事情节完整有趣，作品里充满了小作者探索海洋深处无限的好奇心，也隐含着浓浓的父爱。在潜水器设计方面，作品体现了作者的想象力和创造力，原汁原味的手绘以及汉字和拼音的组合，彰显了孩子的童真。

## Expert Comments:

The story in the book *Little Flower Treasure Hunt* is complete and interesting. It is full of the young author's infinite curiosity to explore the deep sea, and also implies a strong father's love. The submersible design reflects the author's broad imagination and creativity. The original hand-painted, Chinese characters and spelling handwriting highlight the childlike innocence of children.

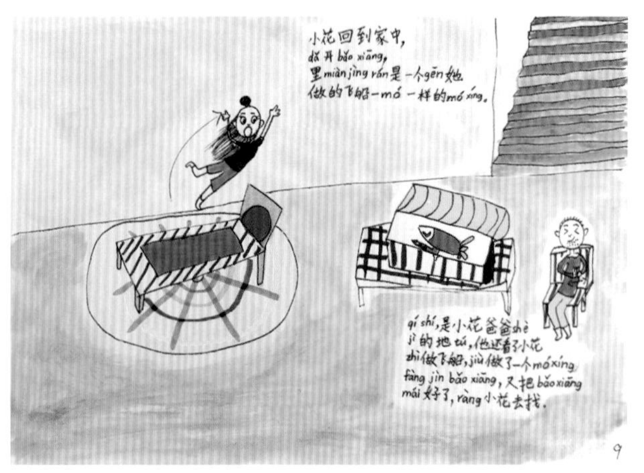

## 《来自远方的礼物》中国海洋大学、青岛科技大学

*A Gift From Far-Away Place*
Ocean University of China, Qingdao University of Science and Technology

大学组
葛昀缇　郑兆丰
University Group
Ge Yunti, Zheng Zhaofeng

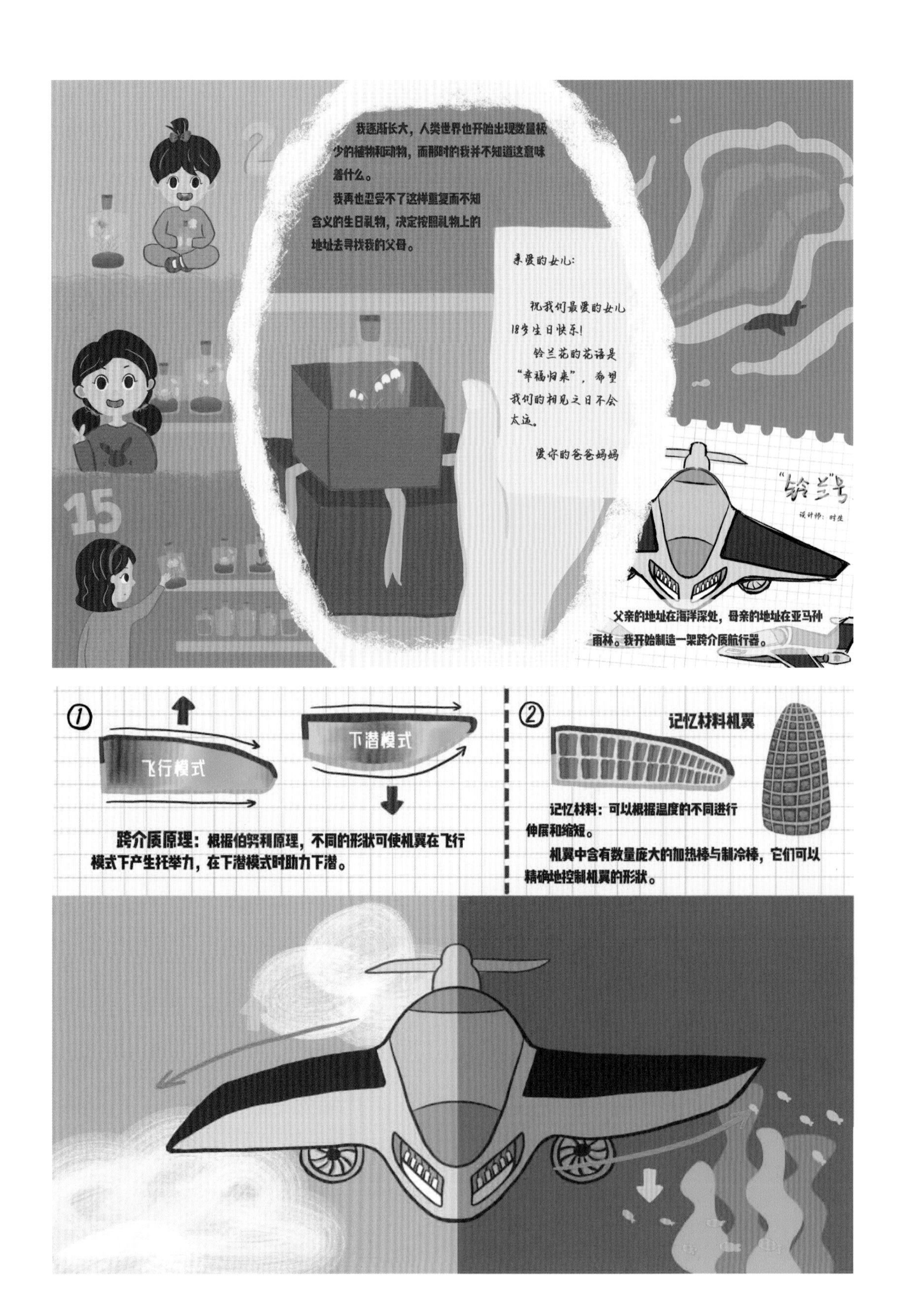
15
我逐渐长大，人类世界也开始出现数量极少的植物和动物，而那时的我并不知道这意味着什么。
我再也忍受不了这样重复而不知含义的生日礼物，决定按照礼物上的地址去寻找我的父母。
亲爱的女儿：
祝我们最爱的女儿18岁生日快乐！
铃兰花的花语是"幸福归来"，希望我们的相见之日不会太远。
爱你的爸爸妈妈
"铃兰号"
设计师：时生
父亲的地址在海洋深处，母亲的地址在亚马孙雨林。我开始制造一架跨介质航行器。
①
飞行模式
下潜模式
跨介质原理：根据伯努利原理，不同的形状可使机翼在飞行模式下产生托举力，在下潜模式时助力下潜。
②
记忆材料机翼
记忆材料：可以根据温度的不同进行伸展和缩短。
机翼中含有数量庞大的加热棒与制冷棒，它们可以精确地控制机翼的形状。

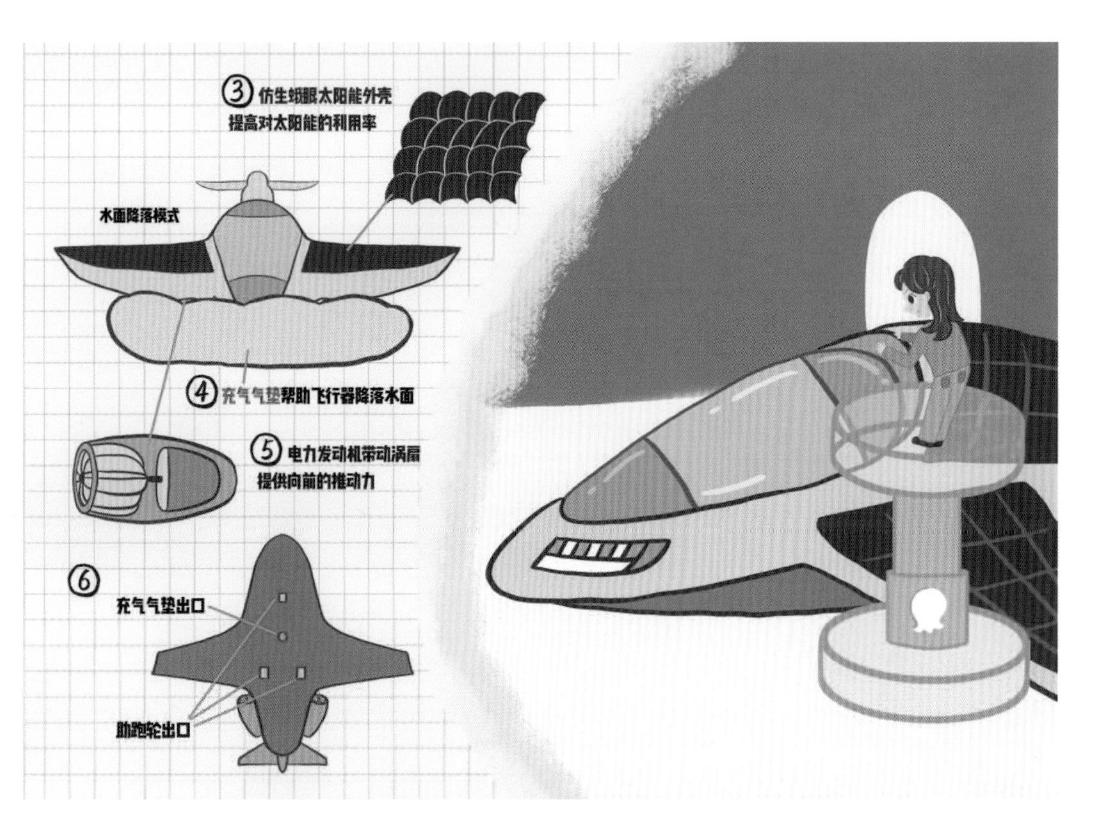
③ 仿生蛾眼太阳能外壳
提高对太阳能的利用率
水面降落模式
④ 充气气垫帮助飞行器降落水面
⑤ 电力发动机带动涡扇
提供向前的推动力
⑥
充气气垫出口
助跑轮出口

我驾驶着航行器飞到了亚马孙雨林的上空，我看到了一个巨大的“玻璃罩子”，里面是我在照片中见过的大片森林，这片森林与周围更加广袤的荒地形成了鲜明对比。

研究室
我见到了母亲，她告诉我，每年送给我的
礼物都是她在这个巨大温室中“复活”的物种。

见过母亲后，我对父亲的礼物和他的工作更加好奇。
我驾驶着航行器来到父亲所在的小岛。

在小岛上的研究所里，我并未找到父亲。父亲的同事告诉我，我的父亲在很深的海底工作，到那里的潜艇每年只来回一次，而我的航行器所携带的电能并不足以支撑我到达父亲所在的研究室。
父亲的同事帮我改造了“铮兰”号航行器能源系统，使它可以在海底充电站充电。

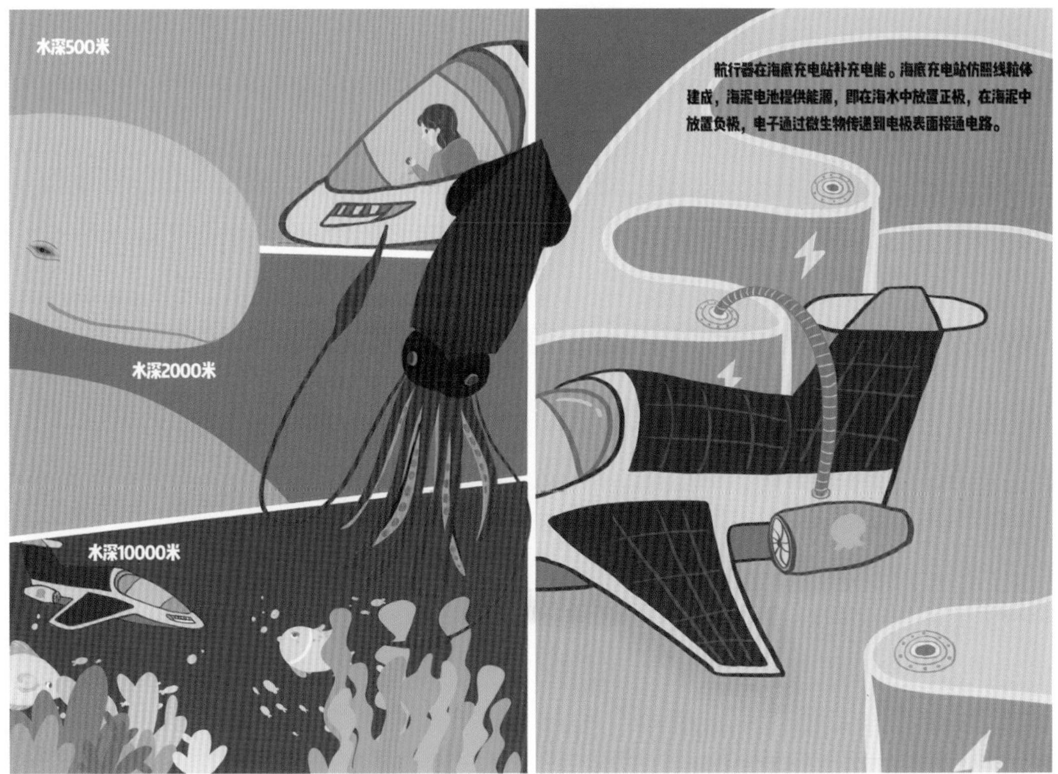
水深500米
水深2000米
水深10000米
航行器在海底充电站补充电能。海底充电站仿照线粒体建成，海泥电池提供能源，即在海水中放置正极，在海泥中放置负极，电子通过微生物传递到电极表面接通电路。

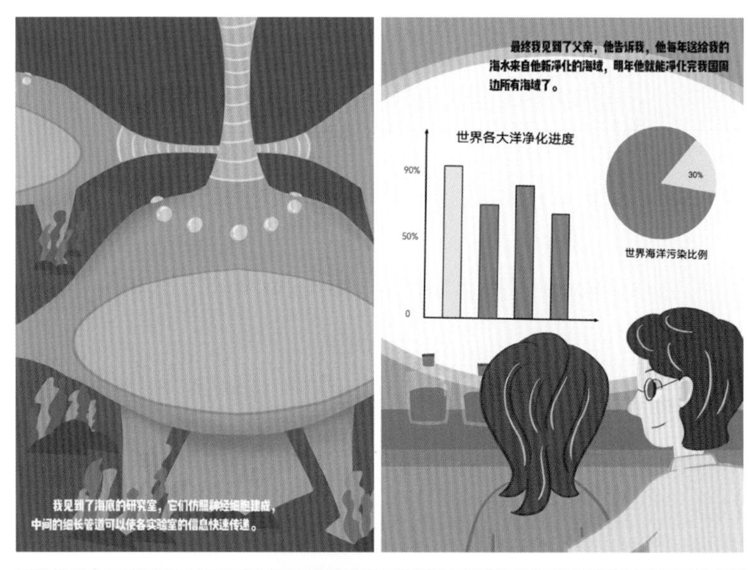

**专家点评：**

《来自远方的礼物》结构安排合理，故事情节生动完整，引人入境。作者选题紧紧围绕生态环境保护，格局大、站位高，胸怀人类可持续发展的主题，符合时代主基调。航行器能源采用绿色设计理念，考虑周全，具有较强的创新性。

## Expert Comments:

The structure arrangement of *A Gift From Far-Away Place* is reasonable. The story is vivid and complete, which grabbed the attention of readers. The authors selected the topic closely around ecological environment protection, which reflects their big pattern and high position. They care about the sustainable development of mankind that is in line with the needs of today's main focus. The vehicle energy uses the green design concept, considers comprehensively, and has strong innovation.

《临渊而动——最后的火种》上海海洋大学

*The Last Spark of Fire — Against the Deep*
Shanghai Ocean University

大学组
牛文卉
University Group
Niu Wenhui

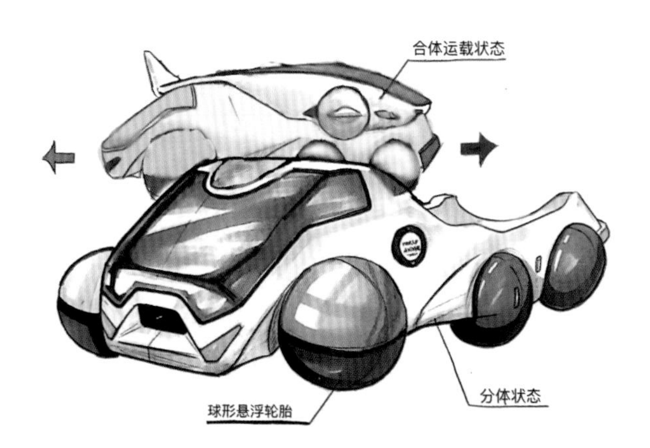
合体运载状态
球形悬浮轮胎
分体状态

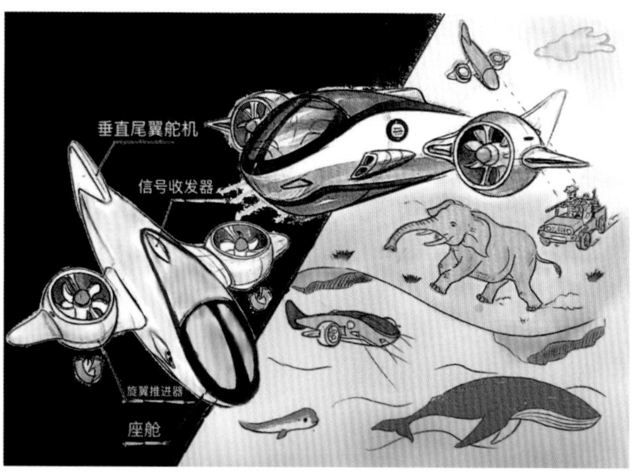
垂直尾翼舵机
信号收发器
旋翼推进器
座舱

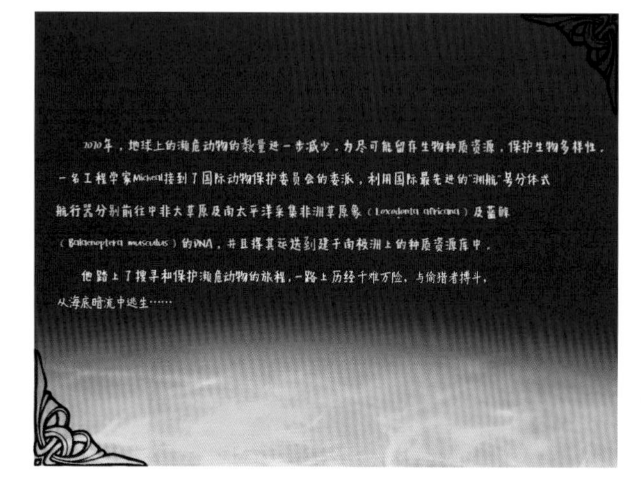
2070年，地球上的濒危动物的数量进一步减少，为尽可能留存生物种质资源，保护生物多样性，一名工程学家Micheal接到了国际动物保护委员会的委派，利用国际最先进的"渊航"号分体式航行器分别前往中非大草原及南太平洋采集非洲草原象（Loxodonta africana）及蓝鲸（Balaenoptera musculus）的DNA，并且将其运送到建于南极洲上的种质资源库中。
他踏上了搜寻和保护濒危动物的旅程，一路上历经千难万险，与偷猎者搏斗，从海底暗流中逃生……

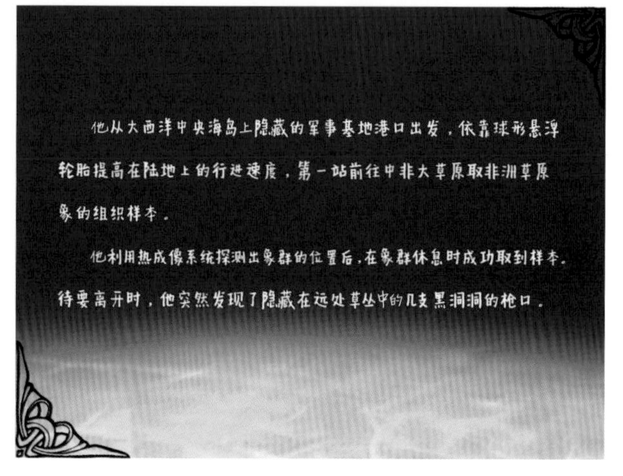
他从大西洋中央海岛上隐藏的军事基地港口出发，依靠球形悬浮轮胎提高在陆地上的行进速度，第一站前往中非大草原取非洲草原象的组织样本。
他利用热成像系统探测出象群的位置后，在象群休息时成功取到样本。待要离开时，他突然发现了隐藏在远处草丛中的几支黑洞洞的枪口。

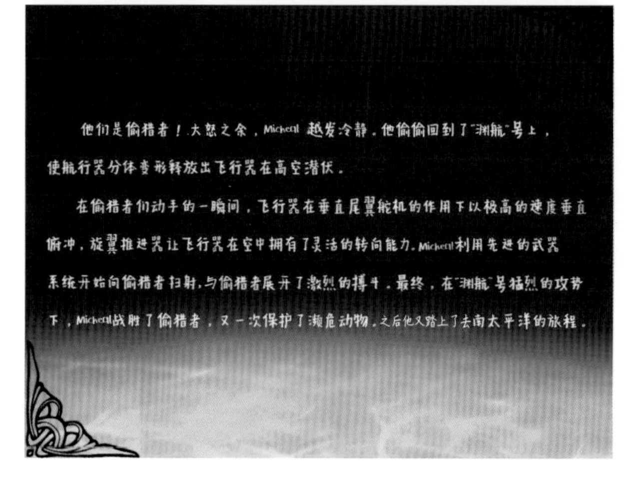

**专家点评：**

《临渊而动——最后的火种》故事情节生动、结构完整，选题围绕濒危野生动物的保护展开，所设计的航行器成为保护濒危野生动物的卫士，具有积极的意义。

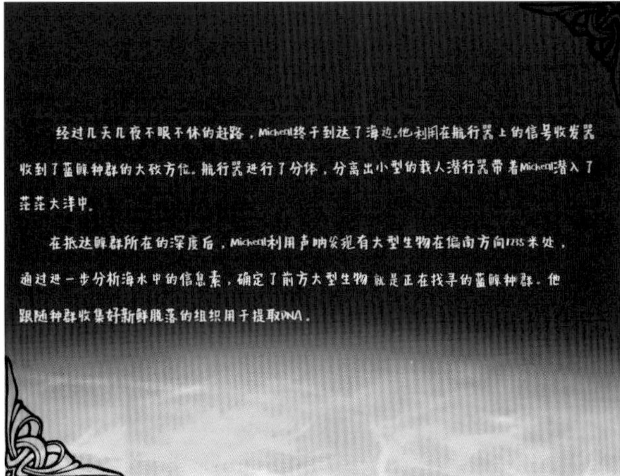
经过几天几夜不眠不休的赶路，Micheal终于到达了海边.他利用在航行器上的信号收发器收到了蓝鲸种群的大致方位.航行器进行了分体，分离出小型的载人潜行器带着Micheal潜入了茫茫大洋中.
在抵达鲸群所在的深度后，Micheal利用声呐发现有大型生物在偏南方向1735米处，通过进一步分析海水中的信息素，确定了前方大型生物就是正在找寻的蓝鲸种群.他跟随种群收集好新鲜脱落的组织用于提取DNA.

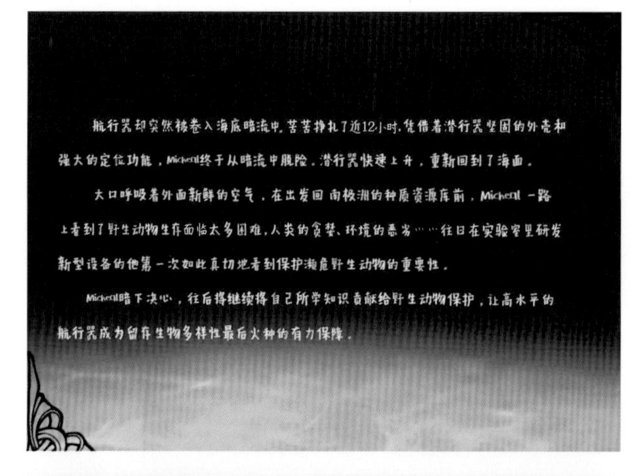

## Expert Comments:

The plot of *The Last Spark of Fire — Against the Deep* is vivid and well structured. The topic focuses on the protection of endangered wild animals. The vehicle design has become a guardian of the protection of endangered wild animals, which has positive significance.

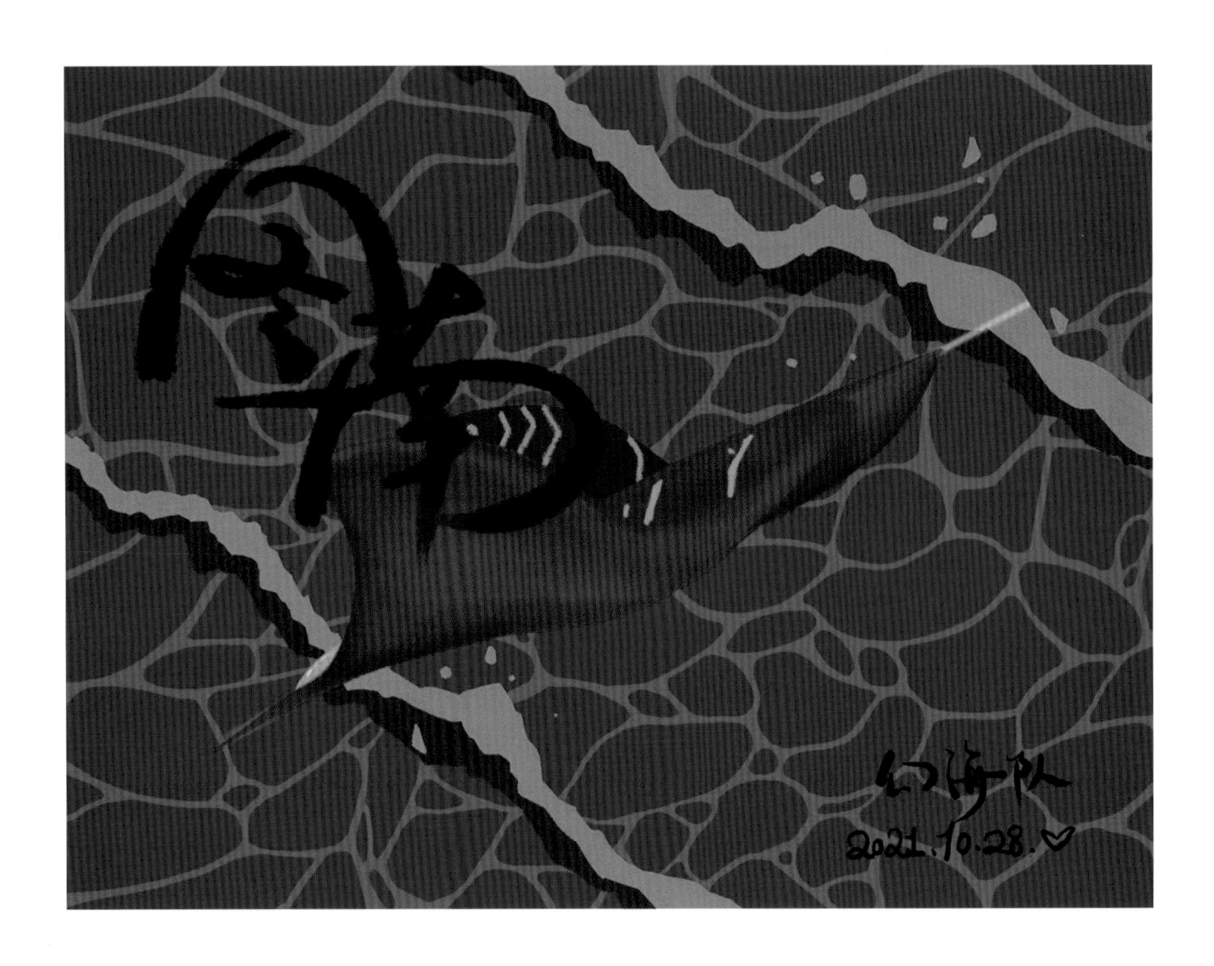


《图南》中国海洋大学

*Tu Nan*
Ocean University of China

大学组
刘海榕　杜柯昀　张梦宇
University Group
Liu Hairong, Du Keyun, Zhang Mengyu

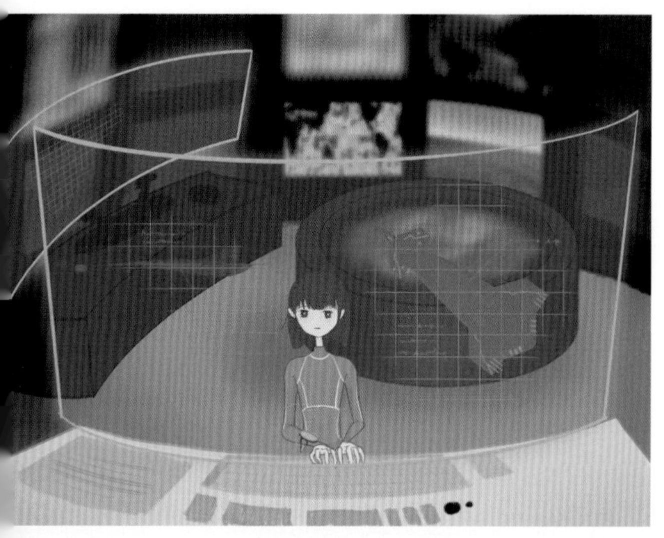

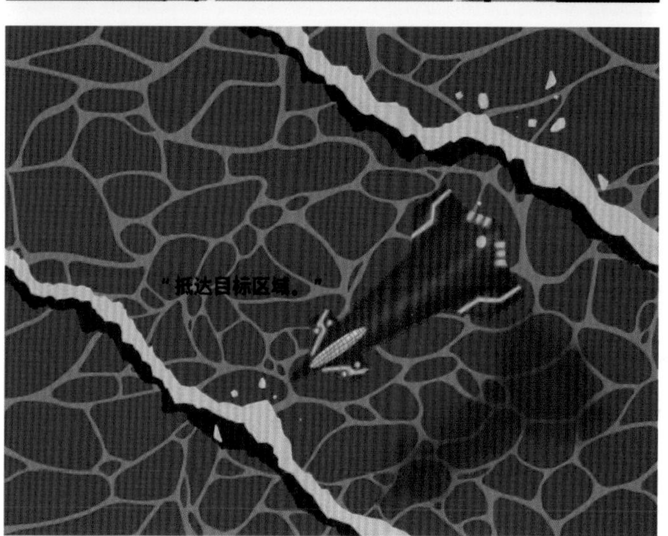


**专家点评：**

《图南》作品结构完整，故事结尾让人意犹未尽。生动的故事情节反映了图南及其母亲代表人类不畏艰险、探索海洋的执着、坚持和付出。潜水器及其探测工具从水下能量转化、变体等角度进行设计，构思新颖、创新性强。

航行器逐渐下降，达到“变体入水”的可行高度后，调整机头，折叠机翼，俯冲入水。

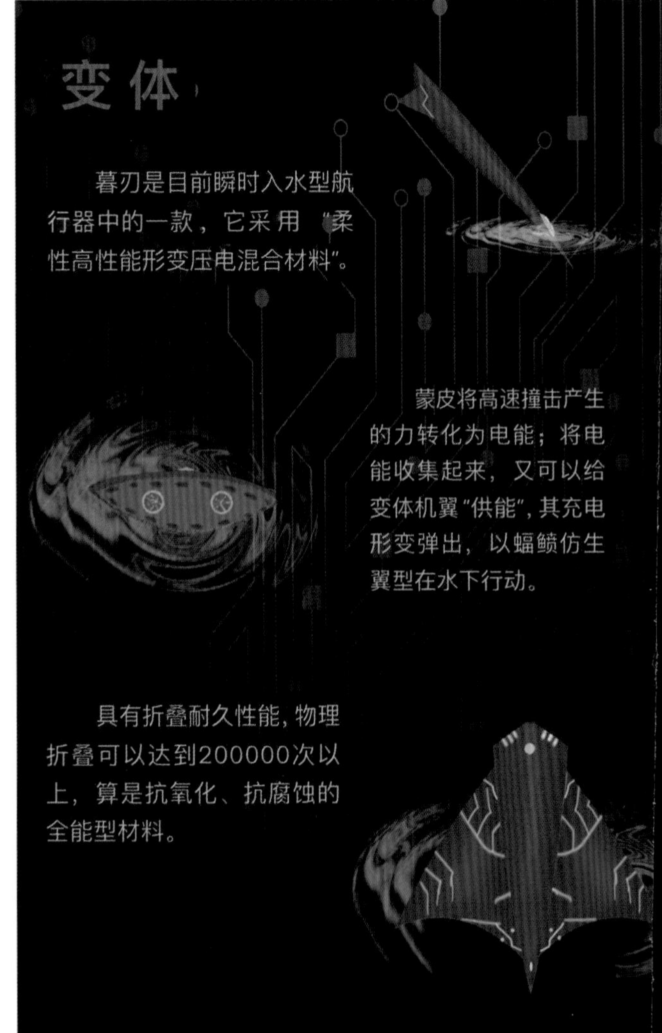
变体
暮刃是目前瞬时入水型航行器中的一款，它采用“柔性高性能形变压电混合材料”。
蒙皮将高速撞击产生的力转化为电能；将电能收集起来，又可以给变体机翼“供能”，其充电形变弹出，以蝠鲼仿生翼型在水下行动。
具有折叠耐久性能，物理折叠可以达到200000次以上，算是抗氧化、抗腐蚀的全能型材料。

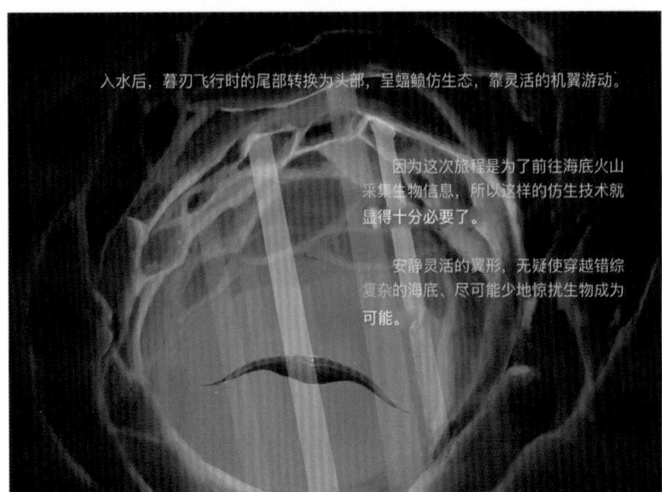
入水后，暮刃飞行时的尾部转换为头部，呈蝠鲼仿生态，靠灵活的机翼游动。
因为这次旅程是为了前往海底火山采集生物信息，所以这样的仿生技术就显得十分必要了。
安静灵活的翼形，无疑使穿越错综复杂的海底、尽可能少地惊扰生物成为可能。

透过窗，能看到绵延的海底火山，它们好像安睡的巨龙。谁能想到20年前这里曾有过一场十分罕见的火山喷发，在这场灾难中，图南的妈妈驾驶的NB2000航行器失事坠毁在爆炸和蒸汽里。

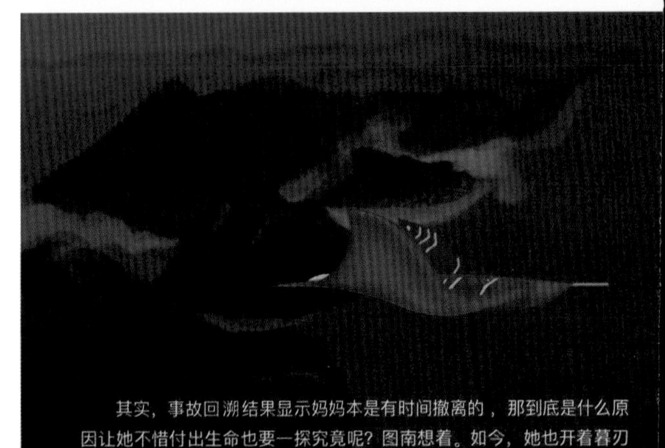
其实，事故回溯结果显示妈妈本是有时间撤离的，那到底是什么原因让她不惜付出生命也要一探究竟呢？图南想着。如今，她也开着暮刃来到这片海域了。

**NB2000 到 NB5000，MTV(Manned Transmedium Vehicle)** 发展了20年，那时候的跨介质载人航行器还处于发展阶段，而现在的暮刃是科技爆炸后的产物——

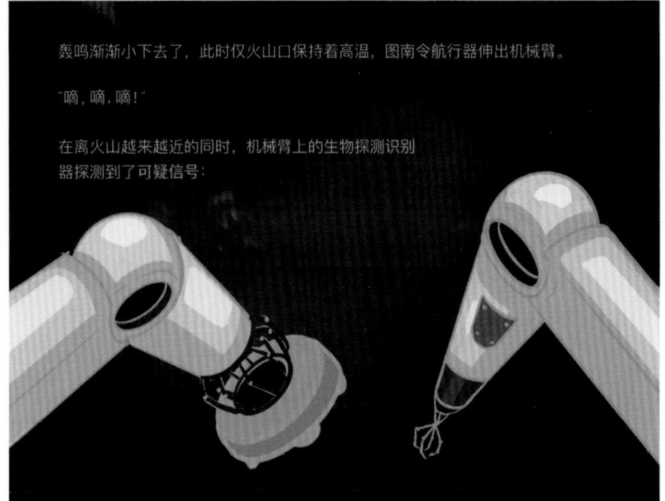


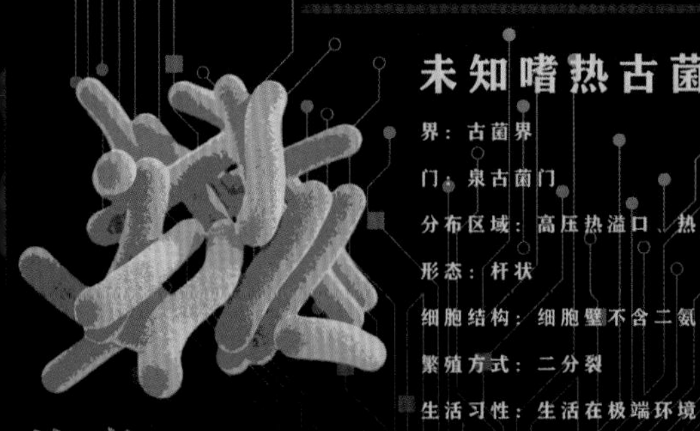

**未知嗜热古菌**

界：古菌界
门：泉古菌门
分布区域：高压热溢口、热泉等
形态：杆状
细胞结构：细胞壁不含二氨基庚二酸
繁殖方式：二分裂
生活习性：生活在极端环境

**注意：**

**可分泌能辅助生产者固氮的物质，死亡后可被分解为效用等同于生长素的神秘物质，可大幅提升生产者的生产能力。**

她突然意识到，这很可能就是妈妈为之付出生命的东西。

这种菌型只在30多年前的西太平洋海域被少量检测提取到，由于样本过少和种群培养的失败而未能被进一步研究利用。

而20年前那次火山喷发，让休眠的芽孢再次被唤醒，探测器检测到了珍贵的信号，只是不幸的是，爆炸式的火山喷发让一切都永远葬身在这片海域……

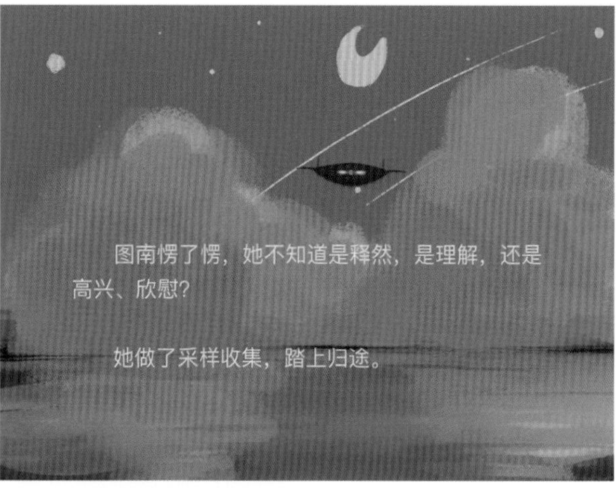

## Expert Comments:

The structure of *Tu Nan* is complete and the end of the story leaves you wanting more. The vivid plot reflects the persistence and dedication of Tu Nan and her mother to explore the ocean on behalf of humanity. The submersible and its detection tools are designed from the perspective of underwater energy transformation and so on. The idea is novel and innovative.

# 设计类竞赛规则及特等奖作品

Competition Rules and Grand Prize Works
in Design Category

**1.比赛题目**

任选在空中（此处为天空而非太空）飞行、水面航行与水下潜航中的2种或3种组合，实现跨介质航行的新概念无人航行器。空气和水的物理性质有着很大的差别，水的密度是空气的800多倍，黏性系数是空气的59倍，因此飞行器和潜航器在航行原理、布局、稳定性、操纵性、材料、结构、动力等方面存在较大差异。跨介质航行器的设计需要协调飞行器和潜航器不同的设计要求，并要兼顾飞行状态、潜航状态和水面航行的设计要求。

**2.比赛形式**

新概念航行器创意设计海报与演示。

**3.比赛说明**

1）参赛对象

大学组：在校研究生、本科生、专科生；

中小学组：在校高中生、初中生、小学生。

每组队员不超过5人。

2）参赛作品要求

预赛上传设计图（包含三视图、渲染效果图及说明），A3尺寸1页，JPG文件，RGB格式，分辨率300 dpi。决赛准备路演PPT，如有视频，需为MP4格式，时长不超过2分钟。中小学组可免做渲染效果图。

3）比赛流程

根据国家与学校防疫要求，赛事全程在网上进行，比赛分为预赛和决赛两个阶段。

预赛阶段，采取专家网评形式，参赛者提交作品电子版（手绘作品为扫描版），择优进入决赛。

决赛阶段，入围作品进行线上路演，每支队伍路演与专家问答时间不超过15分钟。

**4.比赛规则**

比赛采取评分形式，得分高者获胜。

满分为100分，具体评分标准如下。

（1）要素的设计（40分）：应包含适合两种或三种情境的航行器结构设计（0分或10分）、动力系统（0分或10分）、导航定位系统（0分或10分）、环境感知系统（0分或10分）。

（2）跨介质的技术创意设计（1~30分）。

（3）航行器材料设计（1~10分）。

（4）航行器工艺设计（1~10分）。

（5）智能化的创意（1~10分）。

对专家的评分进行统计，去掉一个最高分，去掉一个最低分，其余评分取平均值，作为该作品的最终得分。

**5.设计参考**

（1）重量设计：将跨介质飞行器设计得足够轻，虽可满足飞行的需求，但会使其下潜变得非常困难。在跨介质飞行器漂浮于水面时，目前有两种方法可使其下潜、上浮：一种方法是改变飞行器的密度，即减小飞行器的排水量或者增加重量；另一种方法是利用机翼、尾翼等升力面产生向下的负升力，通过改变负升力的大小来实现下潜、上浮。

（2）外形设计：如飞行器不改变外形，直接进入潜航状态，则机翼在水中仍会产生向上的升力。要解决这一矛盾，只能调整潜航时的姿态，使机翼的迎角减小至零或使机翼产生负升力，但这必然导致水下航行的阻力大大增加。变体技术是解决这一矛盾的有效手段。同时，采用变体技术也可解决海基和空基发射平台空间有限、跨介质飞行器的尺寸受到严格限制的问题。

（3）其他设计难点：跨介质的基本原理、能源动力、通信导航、复合材料、变体技术、发射技术、控制技术等。

（4）相关跨介质航行器资料可查阅但不限于大赛网站。

注：本规则的解释权归大赛组委会所有。

## 1.Description

Design a new concept unmanned vehicle which can choose two or a combination of three kinds of flight in the air, surface navigation and underwater navigation to achieve -media navigation. Because the physical properties of air and water are very different, the density of water is more than 800 times that of air, and the viscosity coefficient is 59 times that of air, so there are great differences between aircraft and underwater vehicle in navigation principle, layout, stability, maneuvering, material, structure, power and other aspects. The design of trans-media vehicle needs to coordinate different design requirements of aircraft and underwater vehicle, and also needs to take into account the design requirements of flight state or underwater state or surface navigation.

## 2.Form

Poster and demonstration of creative design of new concept vehicle.

## 3.More Information

1) Participants

University group: graduate students, undergraduates, and junior college students;

Primary school and high school group: primary school and high school students.

There are no more than 5 players in each group.

2) Requirements for works

Upload the design drawing (including three views, rendering effect drawing and description) for the preliminary competition, the design format is one page in A3 size, JPG file, RGB format, resolution 300 dpi. Final preparation road show PPT. Video files, if available, must be in MP4 format and the duration should not exceed 2 minutes. Primary school and high school students are exempt from rendering renderings.

3) Procedure

According to the requirements of the state and the school, the whole competition was conducted online, and the competition was divided into two stages: preliminary and final.

In the preliminary stage, online evaluation by experts will be adopted. Participants will submit electronic versions of their works (hand-painted scanned versions), and the best works will enter the final.

In the final stage, the finalists will go through an online roadshow, and each team will have 15 minutes to complete the roadshow and Q&A with experts.

## 4.Rules

Scoring mechanism will be used.

The score is 100 points and scoring rules are as follow:

(1) Elements(40 points): The work of Marine vehicles should include two or three scenarios of vehicle structure design (0 or 10 points), power system (0 or 10 points), navigation and positioning system (0 or 10 points), environment sensing system (0 or 10 points).

(2) Trans-media technical creative design (1 to 30 points).

(3) Navigation material design (1 to 10 points).

(4) Craft design of vehicle (1 to 10 points).

(5) The intelligent idea, 1 to 10 points.

The highest score and the lowest score will be removed, then the total score of the referee and the average value will be counted as the score of the work.

## 5.Design Reference

(1) Weight design: Design the trans-media vehicle to be light enough to fly, but it will be very difficult to dive. In the case of the trans-media vehicle floating on the water surface, there are currently two methods to achieve diving and floating:

One way is to change the density of the vehicle, by reducing its displacement or increasing its weight. Another method is to change the size of negative lift to achieve diving and floating, such as the use of wings, tail and other lift surface to generate negative lift downward.

(2) Shape design: if the aircraft does not change shape, directly into the underwater state, then the wing in the water will still produce upward lift. In order to solve this contradiction, we can only adjust the attitude during diving, so that the angle of attack of the wing can be reduced to zero or the wing can produce negative lift, but this will inevitably lead to a great increase in the resistance during underwater navigation. Variant technology is an effective means to solve this contradiction. At the same time, using variant technology can also solve the problems of limited space of sea and space-based launch platforms and strict limitation of the size of trans-media vehicles.

(3) Other design difficulties: basic principle of trans-media, energy and power, communication and navigation, composite materials, variant technology, launch technology, control technology, etc.

(4) Information on trans-media vehicles is available on, but not limited to, the competition website.

Note: The organizing committee reserves the right to interpret the rules.

# 跨介质四栖航行器

## （娱 翔 潜 底）

一、三维图

(a) 三维效果图

(b) 部件组成图

二、三视图

(a) 正视图

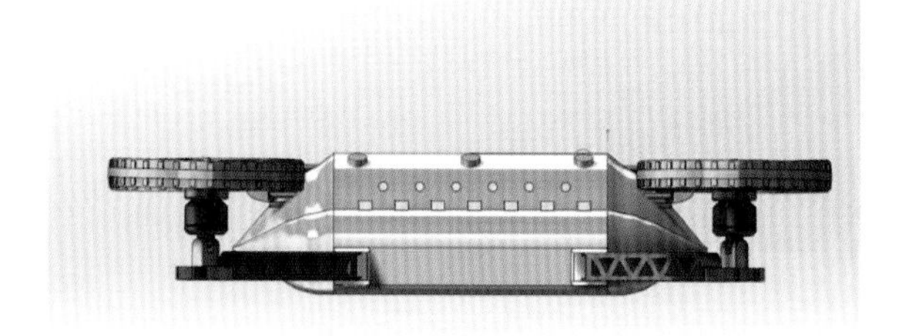

(b) 侧视图

三、工作模式

(a)空中飞行模式

(b)水面船行模式

(c)水下潜航模式

四、跨介质四栖航行器简介

①该跨介质四栖航行器，能够主动识别介质（空气、水、土），并自动切换工作模式；通过航行器两侧压载舱与中空碳纤维杆调整自身重量
航行器结构上的变形功能，最后结合叶片的不同转速（快速、中速、慢速等）与推力，来完成在不同介质中的平稳航行与合理的机械运动。

②在水中或空气中，能够识别潜水员或操作员的手势，实现伴游、辅助拍摄以及安全预警等功能。

③在空气、水跨介质分界面工作时，利用两侧压载舱与中空碳纤维杆，进行平稳的吸排水来缓慢改变密度，同时调整叶片转速，实现稳定起飞

④在空中利用中国北斗导航，在水中采用视觉导航；动力系统采用锂电池电力驱动，结构变形采用水液压驱动。

## 《娱翔潜底》山东省青岛第三十九中学

*Yu Xiang Qian Di*
Qingdao No.39 Middle School

**中小学组**
**张清元　陈叶格　郑韩南丁　王紫琪　杨铭琦**
Primary School and High School Group
Zhang Qingyuan, Chen Yege, Zheng Hannanding,
Wang Ziqi, Yang Mingqi

(c)俯视图

)泥面车行模式　　(e)海底取物模式

度，完成在空气、水等介质中的变重，利用水液压技术完成

潜。

### 专家点评：

该作品针对飞行航拍、浮潜等休闲娱乐场景设计出了一款轻量级、可跨介质的航行器。该航行器整体技术方案高度可行，动力、导航、感知等各个子系统设计科学合理，尤其是传动和动力推进装置体现出了良好的创新性。该作品具有面向消费电子市场孵化出新型产品的潜力。

### Expert Comments:

This work provides a lightweight trans-media vehicle for flying aerial photography, snorkeling and other recreational scenes. The overall technical scheme of the vehicle is highly feasible, the design of the subsystems such as power navigation and perception is scientific and reasonable, especially the transmission and power propulsion devices reflect good innovation. The work has the potential to incubate new products for the consumer electronics market.

# 分离式跨介质航行器
# ——“龙宫”号


队长：储弘锐
队员：赵佳宁、谢文晔、杨博研、刘越
指导教师：韩俊豪



北京师范大学南山附属学校
Nanshan Affiliated School of Beijing Normal University


## 工作模式二：分离式状态

该模式下，“龙宫”号浮至水面后升降平台将驾驶舱升至航行器顶部，驾驶舱展开机翼后起飞，独立执行飞行任务。驾驶舱分离后航行器将智能追踪飞行器，飞行器能源不足时自动返回航行器补充能源。

展开的太阳能板

左视图

主视图

俯视图

折叠式机翼
在航行器内部为向内折叠状态，起飞时向外展开

氢能源动力电池
航行器的主动力来源

GPS
飞行器在空中精准定位

IMU
飞行器飞行姿态解算

交互式接口
用于飞行器和航行器之间信号和能源的交互

折叠的太阳能板
机器浮至水面后可展开充电

蓄电池
飞行器的动力来源

储氧罐
负责飞行器分离后驾驶舱的氧气供给

雷达
负责飞行器飞行时的空中障碍感知

副控系统
负责飞行器分离后航行器的控制和驾驶

电磁铁
位于飞行器底部
用于飞行器和航行器之间的连接和固定

惯性测量单元
水下自主式导航定位

太阳能备用电池
应急备用电源

高压储氢罐
存储氢气，
用于氢能源发电

浮力材料
用于应急上浮

高压储氧罐

声呐
水下地形图测绘及避障

升降平台
飞行器进出舱体的推动装置

CTD
感知海洋温盐环境

卫星信号接收机
水面航行定位设备

水声通讯机换能器
用于在水下接受和发送指令，
实现水下的远程控制和无人驾驶

舱盖
用于飞行器的进出

水下LED灯阵及摄像头
水下视觉感知

## 工作模式一：一体式状态

该模式下，“龙宫”号为水下航行器+飞行驾驶舱的一体式状态。该状态下，飞行驾驶舱为主控装置，通过交互接口实现与水下航行器之间的信号和能源交互，水下航行器保证海底潜行和海面航行任务的实现。

主视图

左视图

俯视图

## 《“龙宫”号》北京师范大学南山附属学校

*Long Gong*
Nanshan School Attached to Beijing Normal University

**中小学组**
**储弘锐　赵佳宁　谢文晔　杨博研　刘越**
Primary School and High School Group
Chu Hongrui, Zhao Jianing, Xie Wenye, Yang Boyan, Liu Yue

**专家点评：**

该作品应对跨域航行的需求，创新式地提出了一种母船——船载飞行器的解决方案，并充分考虑了节能、环保的要求，创新性地使用了太阳能、燃料电池等清洁型能源，整体系统设计准确合理，外观及展示效果美观，体现出了优秀的设计水平。

Expert Comments:

The work meets the needs of cross-domain navigation, and puts forward some innovative aircraft from a mother ship—a solution for shipborne aircraft. Taking full consideration to the energy conservation and environmental protection requirements, this work uses innovative clean energy such as solar and fuel cells. The overall system design is reasonable with a beautiful appearance and effective displaying skills, which reflects an excellent design level.

# “归去来”号

## 设计思路

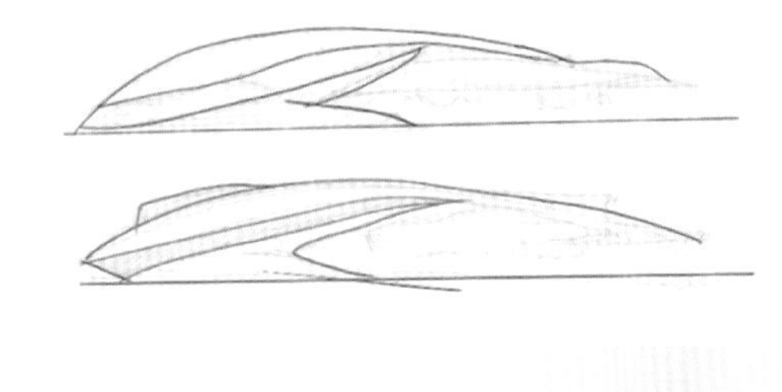

参考鲸鱼身体曲线，经过前期绘制草图，构思创意点，设计航行器，并与无人机结合，最后建模渲染，将之命名为“归去来”号。

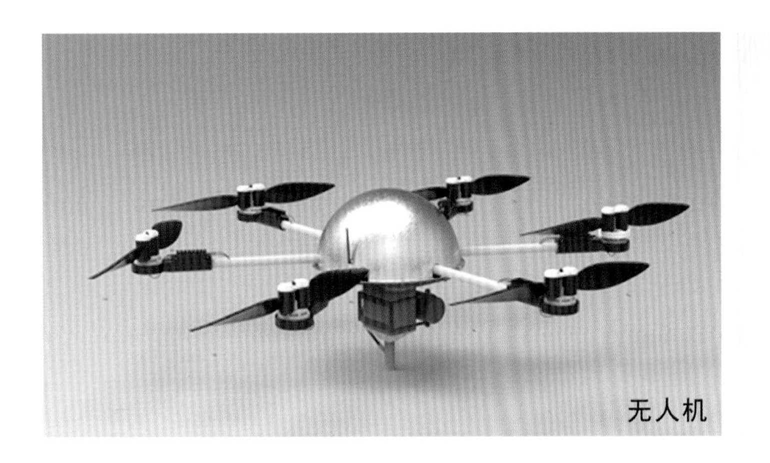
无人机

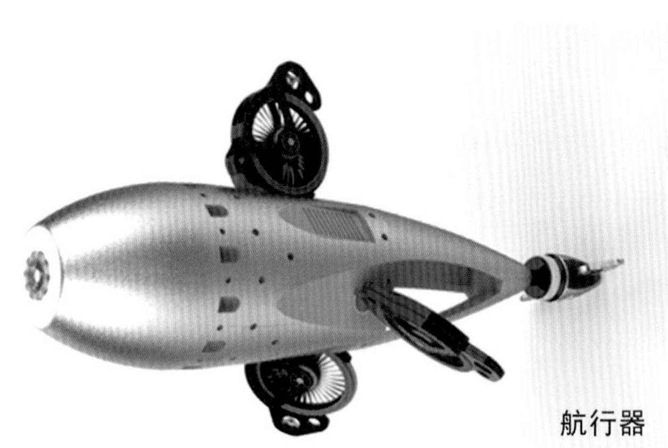
航行器

## 作品原理

处于飞行状态时，机翼旋转为装置提供动力,可上升、下降、前行、后退、急停;入水前折叠顶部的机翼，减少其在水中的阻力；以改变水舱的水量(即改变重力)来控制装置下潜深度。

六个小螺旋桨为空中飞行提供动力，涡轮发动机为水下潜航提供动力。

## 作品创意

1. 环境感知系统

负责环境感知的传感器类似于人的视觉和听觉器官，主要依靠激光雷达、摄像头、毫米波雷达提供的数据进行算法处理。通过传感器、里程计算法，航行器可以感知自身的位置状态，实现自主导航。

2. 航行器材料

混杂复合材料，其抗冲击强度、韧性高，并具有一定的厚度。

3. 航行器工艺

将无人机结构与水下航行器结构设计结合。无人机结构设计满足空中飞行的任务需求；水下航行器结构设计满足水下航行的任务需求。

4. 智能化

完成水下航行与空中飞行任务的装置分离，互不干扰，易于切换航行状态。

5. 导航定位系统

将GPS导航系统和INS（惯性导航）结合，定位精度高，速度快，采用先进的工业级模块和设备，可靠性好。

**《"归去来"号》广东海洋大学**

*Gui Qu Lai*
Guangdong Ocean University

**大学组**
**郎群　郭可盈　邱佳欣　黄硕　鄢子琪**
University Group
Lang Qun, Guo Keying, Qiu Jiaxin, Huang Shuo, Yan Ziqi

**专家点评：**

该作品外观新颖，创意很好，外形、配色等具有科技感和美感；设计符合主题，通过变体实现在空气介质和水介质中的运动，跨介质具有实现的可行性；各分系统的设计有较强的技术可行性，可深入研究和设计，具备可实现性。

Expert Comments:

This work is a good creativity that designs a novel appearance. It has a sense of science & technology and beauty in both appearance and color. The design conforms to the theme, and through the variation of air medium and water medium in two different media movements, cross-media function is feasible. Each subsystem design has strong technical feasibility, in-depth research and design, with realizability.

## 设计说明

- 多航态三栖变体航行器以变体技术、复合式动力系统、多种航行姿态的结构设计为基础实现跨介质航行。
- 通过可折叠式螺旋桨、矢量喷口、可变体机翼、浮潜气囊装置实现了多航态跨介质航行，解决了跨介质航行时存在的重量、体积、浮力、流线性等问题。
- 光敏纤维变色材料、水下耐压设计、GPS及惯性导航系统、航态自动变体等技术，使得航行器有着更好的耐久性、可靠性、完备性，为小型化三栖无人勘测装备的设计提供了新的可能性。

# 多航态三栖变体航行器

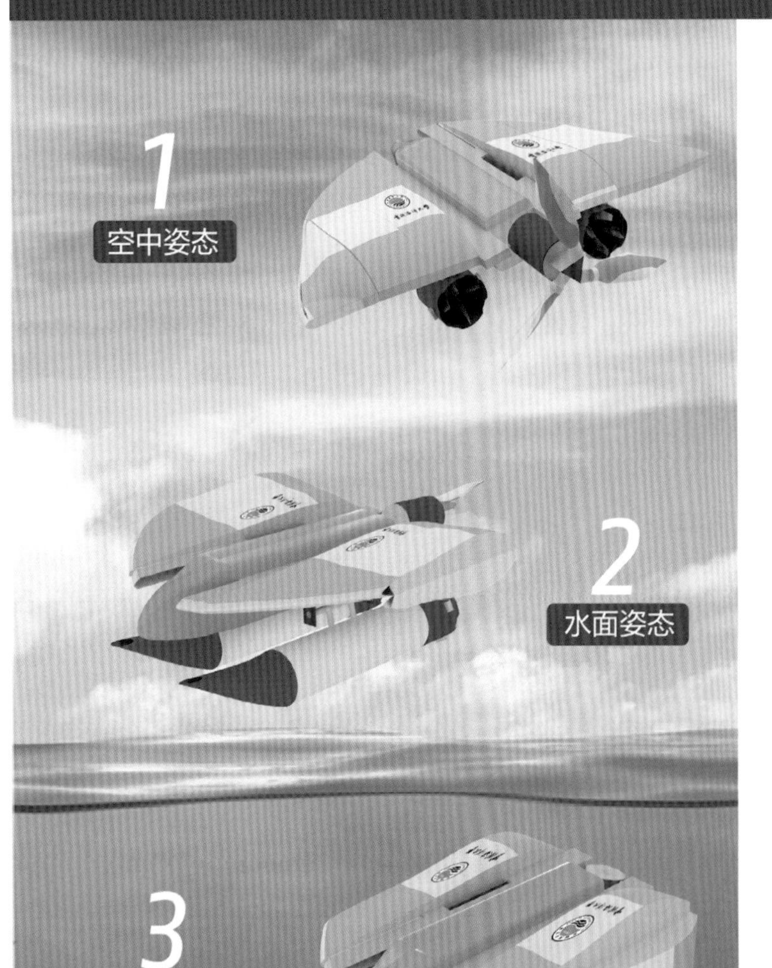


## 跨介质航行

空中姿态：
机翼完全展平，气囊向内折叠，伸缩螺旋桨和矢量喷口提供动力以及转向。

水面姿态：
最外侧机翼内折，上下弧度抵消，减小水面航行时机翼产生的升力，气囊竖直向下折并吸水，使得水面航行时航行器保持稳定。

水下姿态：
整体结构紧凑，表面疏水且具有一定弧度，形成流线体，减小水下阻力。

## 三视图

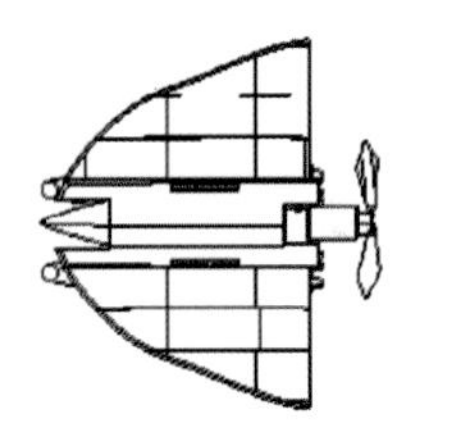

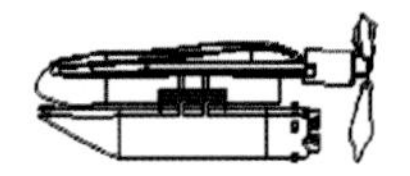

中国海洋大学

王越 王荫东 杨泽祎 葛舟

**《多航态三栖变体航行器》中国海洋大学、青岛农业大学**

*Multimodal Variant Tripod Vehicle*
Ocean University of China, Qingdao Agricultural University

**大学组**
**王越　王荫东　杨泽祎　葛舟**
University Group
Wang Yue, Wang Yindong, Yang Zeyi, Ge Zhou

**专家点评：**

该作品设计完整，符合设计主题要求。感觉它可以如变形金刚一样变体，应用多种新技术实现跨介质飞行。设计者细致考虑了跨介质过程中的状态转换及其可行性，是工程设计中非常好的概念设计。设计者的理论功底深厚、技术扎实，具备深入研究的能力。

Expert Comments:

The design of the work is complete, in line with the design theme requirements. It feels like a transformer and applies a variety of new technologies to achieve cross-media flight. The designers carefully considers the state transformation and its feasibility in the process of cross-media, which is a very good concept design before engineering design. The designers has a profound theoretical foundation, solid technology, and the ability of in-depth research.

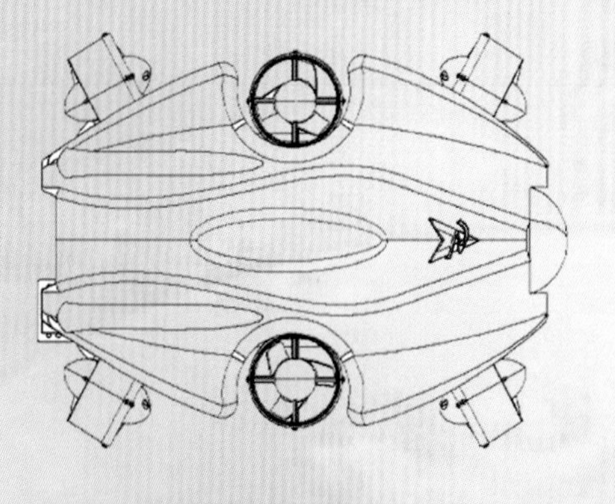

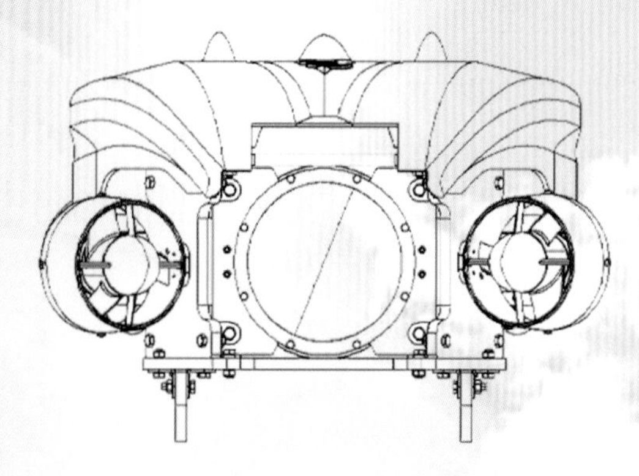

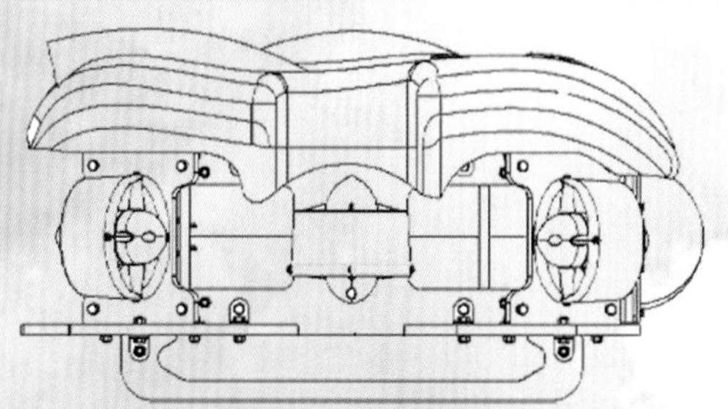

# 『瀚海』号无人航行器

机器人在水下进行工作时，机器人排水的体积$V_{排水}$ 等于航行器整体体积$V$，因此求得航行器的体积 $V$。按照此体积 $V$ 设计机器人的整体体积，以实现水下航行器在水中初始状态“零浮力”和“零重力”；为适应复杂的水下环境，需要有高强度的机身结构以及很好的防水性能，能够下潜 50 米；在水面运动时，中间的推进器保证机器不会因为重量沉入水下，其他4个推进器同时进行推进。

机器人：设计搭载姿态传感器 mpu6050 模块（实时测得机器人在水下的实际姿态，并利用在机器人自身坐标系 3 个坐标轴上的加速度以及角加速度的值，确定机器人的空间位移），深度传感器（使得机器人在水下实时感知水下压力状态，并将其换算为下潜的深度，可使机器人在竖直方向上能够进行闭环控制，准确进行定高悬停等工作状态），酸碱度以及盐度温度等传感器可在机器人执行相关检测任务时进行数据资料 的采集。除此之外，机器人在水下运动还需要搭载图像采集装置、无线图传模块、电机驱动模块等；浮块上专门搭载 GPS 模块，进行水面实时定位。

图像获取处理：我们在机器人前端安装了两个摄像头来获取图像，后台程序自动利用重合度、曝光度等参数确定缝合线具体位置进行图像拼接，实现类似鱼眼大视野的效果，并且利用左右成像的互补信息，优化细节、修补失真并记录航道中各个地方的详细情况。

机器人设计完成后，各部件分别用各种工艺方法进行了加工。

●**浮力方面：**经过对浮力与重力进行计算，运用铸造的方法加工出了配重块并配备浮块，经过实验调节，实现了水中自稳。

●**防腐方面：**机器外壳采用汽车喷漆工艺，对框架进行喷漆处理，螺钉、角码处涂有防水酯，达到防腐效果。

●**水密方面：**控制舱后盖、前罩、开关接头均采用橡胶圈结合防水酯压紧密封，接头处用树脂灌封，并通过抽真空的方式测试控制舱气密性。

●**耐压方面：**控制舱采用铝合金制造，机器整体框架结构紧凑，通过受力分析改善结构，有利于压力的分散。

●**运动方面：**机器采用 6 桨均布的常规动力布局，可以做绕点旋转、斜线平移等动作，且使用四轴传感器能做到运动自稳。

●**减阻方面：**机器外壳设计通过solidworks曲面建模完成，呈流线型。通过motion分析对整台机器进行流体力学分析，并对曲面的5个位置进行镂空处理，以减少阻力。机器框架及电机布局设计一定的斜度，有利于水流流过，结构紧凑，可减少阻力。

**《“瀚海”号无人航行器》郑州大学**

*“Haohan” Unmanned Underwater Vehicle*
Zhengzhou University

**大学组**
**王宇航　余劲明　郑航　姚俊豪　尹豪**
University Group
Wang Yuhang, Yu Jinming, Zheng Hang, Yao Junhao, Yin Hao

**专家点评：**

该作品设计完整，能够实现从水下到水面的运动状态转换，采用双目视觉，提高了观测的逼真度，该作品可以应用于渔业和航道检测，具有较强的实用性。

Expert Comments:

The design of this work is complete, and can realize the motion state conversion from underwater to water surface. It is using binocular vision to improve the fidelity of observation, which can be used in fisheries and waterway detection. It has strong practicality.

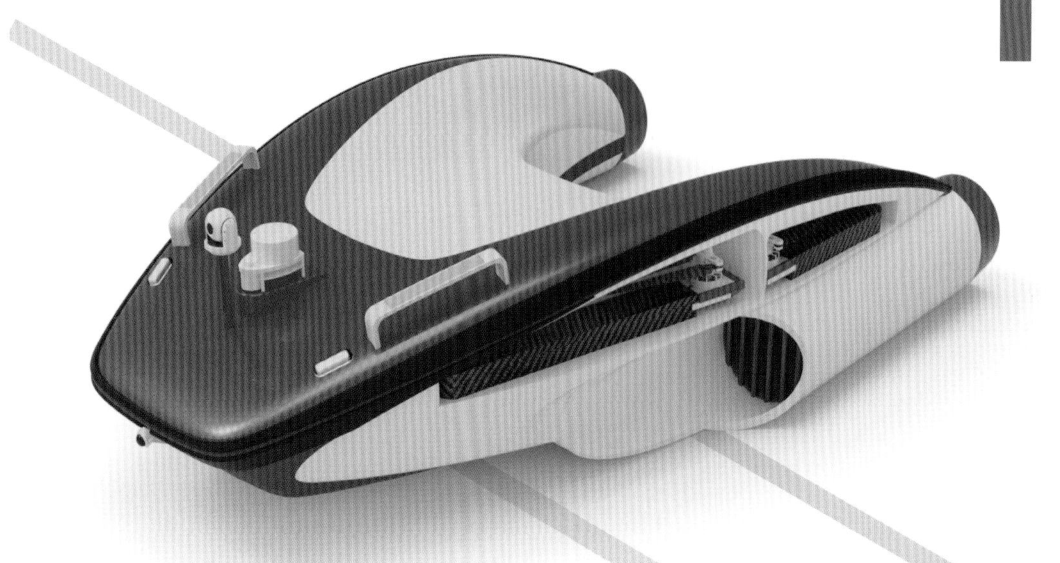

| 主要参数 | |
|---|---|
| 船高 | 0.4米 |
| 设备重量 | 10千克 |
| 水线高度 | 0.2米 |
| 载重量 | 80千克 |
| 航速 | 3米/秒 |
| 飞行速度 | 15米/秒 |
| 续航时间 | 2小时 |
| 工作范围 | 5千米 |

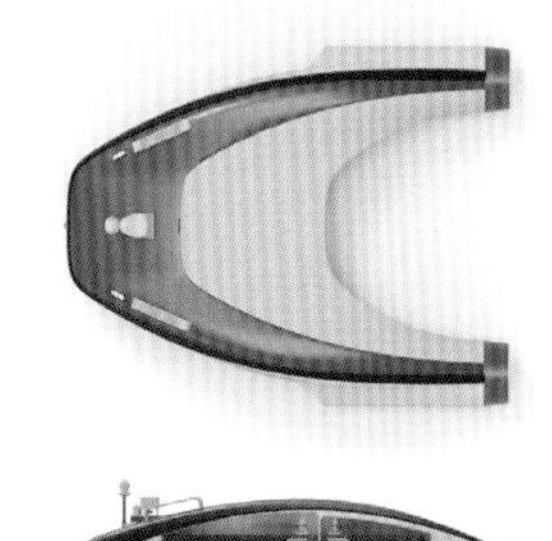

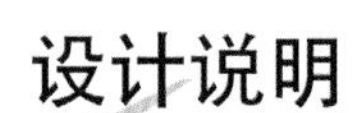

# 智能水空救援装置

## 设计说明

本设备的折叠机翼打开，使其能在空中飞行。飞行过程中通过摄像头获取目标水域的信息，在近海则通过水上人员的动作分析是否有落水者。当发现有落水者时，摄像头锁定该落水者，本装备则迅速飞行到落水者附近进行降落。降落到水面上时，本设备会关闭螺旋桨并驱动电机将螺旋桨折叠至一边，进而将旋翼收至机身中，并在收起机翼的同时驱动喷泵，使本设备行驶至落水者身边停下。落水者抓住机身上的把手，本设备会将该落水者带至岸边。

## 救援过程

巡逻——充电待机

巡逻——扫描识别

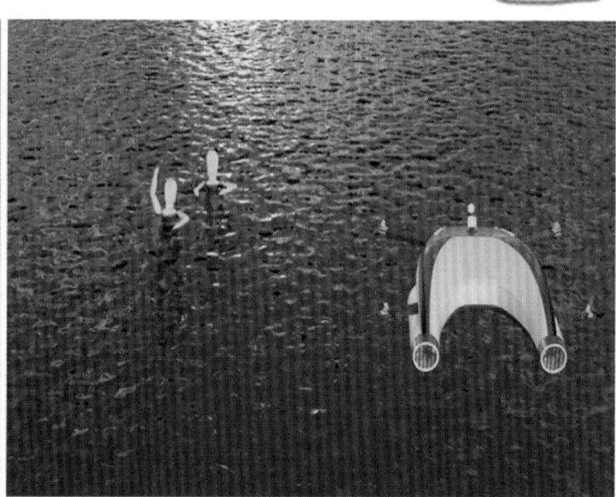

巡逻——目标识别

救援——降落水面

救援——调整姿态

救援——安全返航

**《智能水空救援装置》广东海洋大学**

*Intelligent Water and Air Rescue Device*
Guangdong Ocean University

**大学组**
**翁伟涛　郭振淇　杨诗丹　赖炯龙　邹远停**
University Group
Weng Weitao, Guo Zhenqi, Yang Shidan, Lai Jionglong, Zou Yuanting

**专家点评：**

该作品应用场景明确，具有很强的实用性，结构设计合理，能够满足从空中到水面运动状态转变的要求，外形设计、配色都比较美观，也考虑了在海面上颜色醒目和结构上的实用需求，是一个可以转入产品设计的好的概念设计。

Expert Comments:

The application scene of this work design is clear. It has a strong practicality, and the structure design is reasonable, which can achieve the requirements of motion state change from the air to the water. Both the appearance design and color are pretty, and it also considers in the sea eye-catching and the structure on the practical needs. It is a good concept design that can be transferred to a production design.

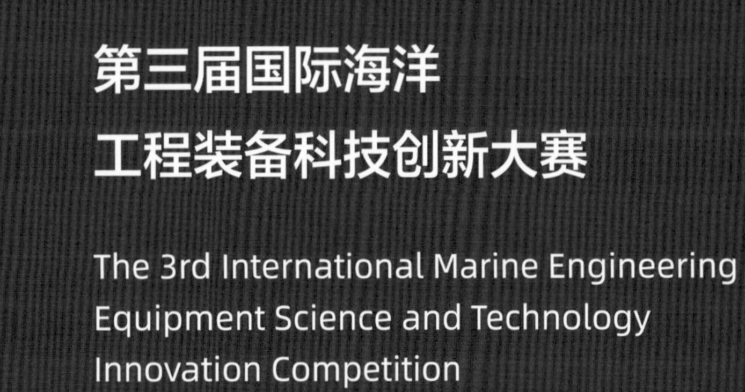

# 第三届国际海洋工程装备科技创新大赛

The 3rd International Marine Engineering Equipment Science and Technology Innovation Competition

# 关于举办第三届国际海洋工程装备科技创新大赛的通知

由教育部高校海洋科学类专业教学指导委员会、教育部高校海洋工程类专业教学指导委员会、中国太平洋学会海洋教育分会、中国海洋大学联合主办的第三届国际海洋工程装备科技创新大赛现已启动。现将有关事项通知如下。

**一、大赛宗旨**

大赛以培养造就国际化海洋科技创新人才为目标，紧密结合海洋工程实际和海洋科学研究，通过海洋工程装备的科技创新设计，引导参赛选手深化对海洋的认识，培养参赛选手的海洋情怀、国际视野、创新精神和实践能力，进而营造全社会关注海洋、了解海洋、热爱海洋、保护海洋、经略海洋的良好氛围。

**二、大赛组织**

主 办 单 位：教育部高校海洋科学类专业教学指导委员会<br>
教育部高校海洋工程类专业教学指导委员会<br>
中国太平洋学会海洋教育分会<br>
中国海洋大学

承 办 单 位：中国海洋大学创新教育实践中心

大 赛 主 席：于志刚（中国海洋大学校长）

大赛副主席：李巍然（中国海洋大学副校长）

## 三、大赛主题

未来海洋交通创想

## 四、比赛项目

本次大赛设科幻类、设计类、仿真类三个类别，科幻类、设计类设大学组、中小学组，仿真类设大学组。

## 五、比赛报名

本次大赛采取网上报名方式，报名以参赛队伍为单位，报名网址：http://47.105.161.112。

报名截止时间：2022年8月21日。

## 六、时间安排

1）初赛时间

科幻类、设计类：2022年9月25日23:59前提交作品与解题说明书；

仿真类：2022年8月15日至9月11日23:59通过仿真平台上传初赛成绩。

2）初赛结果公布时间

科幻类、设计类：2022年9月29日；

仿真类：2022年9月14日。

3）决赛时间

2022年10月中旬，具体时间另行通知。

## 七、奖项设置

大赛设立特等奖、一等奖、二等奖。

## 八、注意事项

（1）受疫情影响，本届大赛全程采用线上评比形式。

（2）科幻类与设计类参赛队伍提交作品时需按要求一并提交解题说明书，初赛网评后公布进入决赛名单，决赛将以线上答辩形式进行。

（3）仿真类无须填写解题说明书，仿真平台将于2022年8月15日在大赛官网开放下载。参赛队伍需自行在仿真平台完成任务并上传初赛成绩。进入决赛的队伍需在规定时间内将可执行exe文件打包发送至大赛邮箱gjhygczb@163.com，决赛将以线上直播形式进行。

（4）如有涉密项目，请作脱密处理后再行申报。

（5）各升学阶段学生均可报名，以决赛时间所读院校为准。

（6）竞赛规则详见《第三届国际海洋工程装备科技创新大赛竞赛规则》。

国际海洋工程装备科技创新大赛组委会

2022年6月22日

# Notice on the 3rd International Marine Engineering Equipment Science and Technology Innovation Competition

The 3rd International Marine Engineering Equipment Science and Technology Innovation Competition co-hosted by Teaching Guiding Committee of Marine Science Specialty in Colleges and Universities of the Ministry of Education, Teaching Guiding Committee of Marine Engineering Specialty in Colleges and Universities of the Ministry of Education, Ocean Education Branch of China Pacific Society and Ocean University of China, started to register now. Hereby notify the related matters as follows.

## 1. Competition Value

The value of this competition is to cultivate international innovative talents in marine science and technology, closely integrate the reality of marine engineering and marine science research. To inspire participants to deepen the understanding of the ocean through science and technology innovation of marine engineering equipment, and cultivate their ocean sensations, international visions, innovative spirits and practical abilities so as to promote the whole society to pay attention to the ocean, learn about the ocean, understand the ocean, love the ocean, protect the ocean, and develop the ocean.

## 2. Organizing Committee

### Sponsors:

Teaching Guiding Committee of Marine Science Specialty in Colleges and Universities of the Ministry of Education

Teaching Guiding Committee of Marine Engineering Specialty in Colleges and Universities of the Ministry of Education

Ocean Education Branch of China Pacific Society

Ocean University of China

**Organizer:**

Innovation Education and Practice Center of Ocean University of China

**Chair of the Competition:**

Prof. Yu Zhigang (President of Ocean University of China)

Deputy Chair of the Competition:

Prof. Li Weiran (Former Vice-President of Ocean University of China)

## 3. Theme

The Future Marine Transportation Imagination and Creation

## 4. Description

The competition events are divided into three categories: science fiction, design, and simulation. Science fiction and design consist of university group, primary school and high school group. Simulation is for university group only.

## 5. Registration

The competition registration is online registration, and the registration site is "http: //47.105.161.112/". Please register with teams, and one member could make up for a team.

Registration deadline: 21st August, 2022.

## 6. Schedule and Venue

1) Preliminary

Science Fiction, and Design: Please submit entries and descriptions before 23:59 BJT on 25th September, 2022.

Simulation: Please upload the preliminary results via the simulation platform from 15th August to 11th September, 2022.

2) Preliminary Results Announcement

Science Fiction, and Design: 29th September, 2022.

Simulation: 14th September, 2022.

3) The Finals

In mid-October 2022, the specific date will be announced separately.

## 7. Awards

The competition will be endowed with three prizes: the grand prize, the first prize and the second prize.

## 8. Matters Needing Attention

(1) Due to the impact of epidemic, this competition runs online only.

(2) The Science Fiction, and Design teams must submit descriptions (specifications) along with entries. Finalists will be selected on the basis of preliminary competition results, and the final will be conducted in the form of online roadshow.

(3) Simulation teams do not have to write descriptions. The simulation platform will be open for download on the official website of the competition on 15th August, 2022. Teams are required to complete the task on the simulation platform and upload the preliminary results. The teams that make it to the finals need to package and send the executable (exe.) files to gjhygczb@163.com within the allotted time. The final will be conducted in the form of online live streaming.

(4) If there are classified items, please register after the decryption.

(5) Students at all stages of further education can apply, subject to the institutions they attend at the final time.

(6) Please refer to *The 3rd International Marine Engineering Equipment Science and Technology Innovation Competition Rules* for details.

International Marine Engineering Equipment Science and Technology Innovation Competition Organizing Committee

20th June, 2022

**大赛主题：未来海洋交通创想**

道路延伸了人类的足迹，拓展了探索的边界，然而覆盖了地球表面积七成以上的海洋中却没有一条属于人类的“海洋航路”。由于深海环境、水体运动和水下运动物体等都对海洋航行有较大的影响和限制，因此我们畅想的“天高任鸟飞、海阔凭鱼跃”仍未能在人类世界真正实现。如今，随着科技的高速发展与创新能力的提升，人类对深海世界的求知欲与探索欲也在日益增强。就让我们运用创新思维，在科学推动故事情节、科技带动人文情怀中遇见科学的未来，如果某种科学技术得以实现，那么未来海洋将……因此本届大赛将以未来海洋交通创想为主题，开展海洋交通相关的科幻绘本、概念设计与虚拟仿真任务，进行竞技。

Competition Theme: The Future Marine Transportation Imagination and Creation.

Roads extend the boundaries of human footprint and exploration, however, there is no “marine road” that belongs to human beings in the ocean, which covers more than 70% of the earth. Because the deep-sea environment, water movement and underwater moving objects all exert greater impacts and restrictions on marine navigation, “being able to fly through the air like birds, swimming underwater like fish” has not yet been truly realized in the human world. Nowadays, with the development of science and technology as well as the improvement of innovation ability, the desire of human beings for knowledge and exploration of the deep-sea world is increasingly strong as well. Let’s think ahead, to meet the future of science in science promoting the storyline, science and technology mobilizing humanities. If a certain science and technology could be realized, then the marine future will (be like)... Therefore, this competition is themed by “Marine Transportation Imagination and Creation”, and events are science fiction picture books, concept design, and virtual simulation related to marine transportation.

# 科幻类竞赛规则

## Competition Rules in Science Fiction Category

**1.比赛题目**

想象在未来的2070年建设一条水下公路，设想的特定载人海洋航行器能够在该水下公路上航行。构想一种或几种原理方案，实现载人海洋航行器在水下公路上航行时进行定位，以顺利从A点到达B点为故事底层逻辑，且主角必须为人类，不得是拟人化主角。构想两种应用于水下公路的交通装置或设施，使得载人海洋航行器的航行更加便利。

水下公路的路段设定要求包含一段或全程处于距离水面和距离海底都较远的深海路段，水下公路主要路段应为水中悬浮形式，不可采用海底隧道，且航行定位不能采用现有的成熟水下航行技术方案。技术方案设计中可运用原理已知但目前技术还没有实现的设想，也可以采用颠覆性原理。

**2.比赛形式**

硬科幻绘本

注：硬科幻是以科学或科学猜想推动故事情节的科幻作品。

**3.比赛说明**

1）参赛对象

大学组：在校研究生、本科生、专科生。

中小学组：在校高中生、初中生、小学生。

每组成员（包含队长与队员）不超过4人。

2）参赛作品要求

围绕大赛主题，以未来水下公路的航行为故事背景提交硬科幻绘本作品，形式为手绘或计算机绘图，字数、篇幅不限，上传文件为PDF格式，文件大小不超过15M。

3）比赛流程

根据国家与学校防疫要求，赛事全程在网上进行，比赛分为预赛和决赛两个阶段。

预赛阶段，采取专家网评形式，参赛者提交作品电子版（手绘作品为扫描版），作品择优进入决赛。

决赛阶段，入围作品进行线上路演，每支队伍路演与专家问答时间不超过10分钟。

**4.比赛规则**

比赛采取评分形式，得分高者获胜。

满分为100分，具体评分标准如下。

（1）要素的准确性（30分）：公路起点和终点（0或5分）、特定水下载人航行器（0或5分）、公路航行位置定位的原理（0或5分）、交通设施1的名称（0或5分）、交通设施2的名称（0或5分）、故事的主人公（0或5分）。

（2）作品的创意（50分）：故事情节的整体创意（1~10分）、水下公路定位原理的科学创意（1~20分）、交通设施1的创意（1~10分）、交通设施2的创意（1~10分）。

（3）作品的文学性（1~10分）。

（4）绘画的艺术性（1~10分）。

（5）深海路段扣分项：采用了当今现有的成熟技术（0~10分）。

## 1.Description

Imagine building an underwater highway in 2070s, envisioning a specific manned marine vehicle capable of navigating on that underwater highway. Conceive of one or several principle schemes to realize the positioning of manned marine vehicles on underwater highways, with the smooth travel from point A to point B as the underlying logic of the story. Besides, the protagonist must be human being, not an anthropomorphic protagonist. Two types of transportation devices or facilities for underwater roads are conceived to facilitate manned marine vehicles navigation.

The section of the underwater highway is required to include a part or a full section in the deep sea that is far away from the water surface and the seabed. The main section of the underwater highway should be submerged-floating.The undersea tunnel could not be used, and the navigation positioning could not follow the existing mature underwater navigation technology scheme. The known principles without technology realization could be adopted in technical solution design, and so as subversive principles.

## 2. Form

To complete a hard sci-fi picture book.

(Note: Hard sci-fi refers to science fiction tales that are developed and driven by science and technology or scientific speculation.)

## 3. More Information

1) Participants

University group: graduate, undergraduates, and junior college students;

Primary school and high school group: primary school and high school students.

There are no more than 4 players (including the team leader) in each group.

2) Requirements for works

Hard sci-fi picture books should be themed on future underwater road navigation. Both freehand drawing and computer drawing picture books have no limited word count or length. Please submit it in PDF file less than15 M.

3) Procedure

According to the requirements of the state and the school, the whole competition was conducted online, and the competition was divided into two stages: preliminary and final.

In the preliminary stage, online evaluation by experts will be adopted. Participants will submit electronic versions of their works (hand-painted scanned versions), and the best works will enter the final.

In the final stage, the finalists will go through an online roadshow, and each team will have 15 minutes to complete the roadshow and Q&A with experts.

## 4.Rules

Scoring mechanism will be used.

The score is 100 points and scoring rules are as follow:

(1) Elements (30 points): the beginning and end of the highway (0 or 5 points), specific underwater manned vehicle (0 or 5 points), the principle of positioning for highway navigation (0 or 5 points), the name of the first traffic facility (0 or 5 points), the name of the second traffic facility (0 or 5 points), the protagonist (0 or 5 points).

(2) Creation (30 points):plot (1 to 10 points), the scientific creativity of underwater highway positioning principle (1 to 20 points), the creativity of the first traffic device (1 to 10 points), the creativity of the second traffic device (1 to 10 points).

(3) Literature (1 to 10 points).

(4) Artistic (1 to 10 points).

(5)Deep-sea section deduction item(1 to 10 points):using existing mature technology.

# Competition Rules in Design Category

### 1.比赛题目(选择B1、B2任一赛题即可)

B1　水下无人航行器

想象在未来的2030年建设一条水下航线，设计一款能够在该水下航线上航行的特定无人海洋航行器及包含航线交通设施的配套场景。立足于现代水下航行场景，构想未来水下航线具体应用场景与所需交通设施，完成新型无人海洋航行器在水下航线上的航行与任务。

水下航线的路段设定要求包含一段或全程处于距离水面和距离海底都较远的深海路段，水下航线主要路段应为水中悬浮形式，不可采用海底隧道，且航行定位不能采用现有的成熟水下航行技术方案。技术方案设计中可运用原理已知但目前技术还没有实现的设想，也可以采用颠覆性原理。

注：水下航线所需交通设施可参考设计交通信号规则、声波传播、通信信标、数据传输、航速测量、航向选择、避障(设计避开其他航行器的航线和动态目标)、路径规划等。

B2　水上新概念无人船

想象在未来的2030年建设一条水上航路，设计一款适合于长期在水面航行的新概念无人船及航路的配套设施。新概念无人船能够满足在指定航路上长期自动航行的需求，并实现能源补给、数据传输等功能。配套场景可以基于现有海上航行场景，也可以构想未来水上航路需求，配套场景中的航路设施包含但并不限于中转站、能源补给设施、航路引导设施、数据传输设施等。

### 2.比赛形式

B1：水下无人航行器及航行场景创意设计效果图与演示。

B2：水上新概念无人船及航行场景创意设计效果图与演示。

### 3.比赛说明

1)参赛对象

大学组：在校研究生、本科生、专科生；

中小学组：在校高中生、初中生、小学生。

每组队员(含队长)不超过4人。

2)参赛作品要求

预赛上传航行器概念设计图4张与海报1张，海报中需包含应用场景设计。提交资料具体要求如下：

每张设计图都为A4尺寸，包含白底三视图3张、白底渲染效果图1张；

A3尺寸海报1张（包含三视图、航行器应用于场景图及说明）。

以上作品均为JPG文件，RGB格式，分辨率300 dpi。决赛准备路演PPT，如有视频，需为MP4格式，时长不超过2分钟。

注：中小学组可免做渲染效果图。

3）比赛流程

根据国家与学校防疫要求，赛事全程在网上进行，比赛分为预赛和决赛两个阶段。

预赛阶段，采取专家网评形式，参赛者提交作品电子版（手绘作品为扫描版），作品择优进入决赛。

决赛阶段，入围作品进行线上路演，每支队伍路演与专家问答时间不超过15分钟。

## 4.比赛规则

比赛采取评分形式，得分高者获胜。

评分满分100分，具体评分标准如下。

（1）要素的准确性（50分）。

B1：作品中应包含水下无人航行器（0或10分）、导航定位系统（0或10分）、动力系统（0或10分）、环境感知系统（0或10分）、水下航线交通设施（0或10分）。

B2：作品中应包含水上新概念无人航行器（0或10分）、数据传输系统（0或10分）、动力及能源补给系统（0或10分）、环境感知系统（0或10分）、水面航路交通设施（0或10分）。

（2）航行器的科技智能化创新设计（1~20分）。

（3）航行器材料工艺创新设计（1~10分）。

（4）航行器造型概念创新设计（1~10分）。

（5）航线场景交通设施创新设计（1~10分）。

## 1. Introduction (please choose either B1 or B2)

B1. Underwater Unmanned Vehicle

Imagine building an underwater route in 2030, and design a specific unmanned marine vehicle capable of sailing along that underwater route with an ancillary scenario containing route traffic facilities. On the basis of the modern underwater navigation, conceive application scenarios of the future underwater route and its required traffic facilities, so as to realize the new unmanned marine vehicle underwater navigation.

The section of the underwater highway is required to include a part or a full section in the deep sea that is far away from the water surface and the seabed. The main section of the underwater highway should be submerged-floating. The undersea tunnel could not be used, and the navigation positioning could not follow the existing mature underwater navigation technology scheme. The known principles without technology realization could be adopted in technical solution design, and so as subversive principles.

Note: The traffic facilities required for underwater routes could refer to the design of traffic signal rules, sound wave propagation, communication beacons, data transmission, speed measurement, course selection, obstacle avoidance (design to avoid other vehicles' route and dynamic target), route planning, etc.

B2. New Concept Water Unmanned Ship

Imagine building a water route in 2030, and design a new concept unmanned ship suitable for long voyage with an ancillary scenario containing route traffic facilities. The new concept of unmanned ship could meet the needs of long automatic voyage on designated water routes, and achieve functions such as energy replenishment, data transmission, etc. The ancillary scenario could be based on the existing offshore navigation, or conceive future water route requirements. Route facilities include but are not limited to transit stations, energy supply facilities, route guidance facilities, data transmission facilities, etc.

## 2. Form

B1. The rendering and demonstration of underwater unmanned vehicle and navigation scenario.

B2. The rendering and demonstration of new concept water unmanned ship and voyage scenario.

## 3. More Information

1) Participants

University group: graduates students, undergraduates, and junior college students;

Primary school and high school group: primary school and high school students.

There are no more than 4 players (including the team leader) in each group.

2) Requirements for works

For the preliminary contest, please upload 4 drawings of vehicle concept design and 1 poster, which must contain the application scenario design. The specific requirements for submission are as follows:

The drawing paper size is A4 each, including 3 drawings of three views on a white background and 1 rendering on a white background.

The poster paper size is A3 (including three views, the vehicle application scenario and description).

The above should be in JPG files, RGB, 300 dpi.

For the final, please prepare the roadshow PowerPoint. Video files, if available, must be in MP4 within 2 minutes.

Note: Renderings are not necessary for primary school and high school group.

3) Procedure

According to the requirements of the state and the school, the whole competition was conducted online, and the competition was divided into two stages: preliminary and final.

In the preliminary stage, online evaluation by experts will be adopted. Participants will submit electronic versions of their works (hand-painted scanned versions), and the best works will enter the final.

In the final stage, the finalists will go through an online roadshow, and each team will have 15 minutes to complete the roadshow and Q&A with experts.

## 4.Rules

Scoring mechanism will be used.

The score is 100 points and scoring rules are as follow:

(1) Elements (50 points).

B1. Underwater unmanned vehicle (0 or 10 points), navigation and positioning systems (0 or 10 points), power system (0 or 10 points), environmental sensing system (0 or 10 points), underwater route traffic facilities (0 or 10 points).

B2. New concept water unmanned vehicle (0 or 10 points), data transmission system (0 or 10 points), power and energy supply systems (0 or 10 points), environmental sensing system (0 or 10 points), water route traffic facilities (0 or 10 points).

(2) The science and technology intelligent innovation design of the vehicle (0 or 20 points).

(3) Vehicle material originality (0 or 10 points).

(4) Vehicle modeling originality (0 or 10 points).

(5) Innovative design of traffic facilities in route scenarios (0 or 10 points).

# Competition Rules in Simulation Category

## 1.比赛题目

随着传统能源日渐枯竭和风力发电技术日趋成熟，风力发电已成为全球新能源的重要来源。中国海上可开发风能资源约7.5亿千瓦，是陆上风能资源的3倍。近年来，我国海上风电产业发展十分迅速，海上风电场数量不断增加、规模不断扩大。但是由于海陆环境差异，海上风电场的迅速发展对日常巡检与维护运行也提出了挑战。因此，为了实现降低经济成本和提高运维效率的目的，赛事围绕海上风电巡检进行仿真类竞赛，在海上风电运维的仿真平台（需在赛事网址下载）上进行编程，通过在指定时间内实现规定动作，完成不同海况下的海上巡检和自主避障等任务。

## 2.比赛形式

按照规则要求进行虚拟仿真航行。

## 3.参赛对象

大学组：在校研究生、本科生、专科生。

每组队员（含队长）不超过4人。

## 4.比赛规则

仿真类初赛与决赛为不同的任务场景。初赛形式为限时完成任务，决赛采取积分制形式，按最终得分高低进行排名。赛事支持用Python、C++、C#、Java语言进行编程控制。

1）初赛规则

赛事预设场景范围为800米×600米，要求在3分钟内巡视风电桩、穿过多岛屿海域并抵达终点，图1为初赛场景，请仔细阅读以下规则。

注：由于赛事需要，场景中距离参数相应缩短。技术参数详见仿真平台操作手册。

海域一：巡视风电桩

参赛选手的无人艇从起点出发，采用自主航行的方式，依次抵达各风电桩附近可识别海域（蓝色区域）并完成相应动作，随后继续往终点方向航行。各风电桩对应动作要求如下。

1号风电桩：南侧通过，过程中不得与风电桩碰撞且距离不得小于1米，绕行过程中与风电桩距离不能大于20米。

2号风电桩：顺时针绕行巡检一周，巡检过程中均在蓝色识别海域内，过程中不得与风电桩碰撞且距离不得小于1米，绕行过程中与风电桩距离不能大于20米。

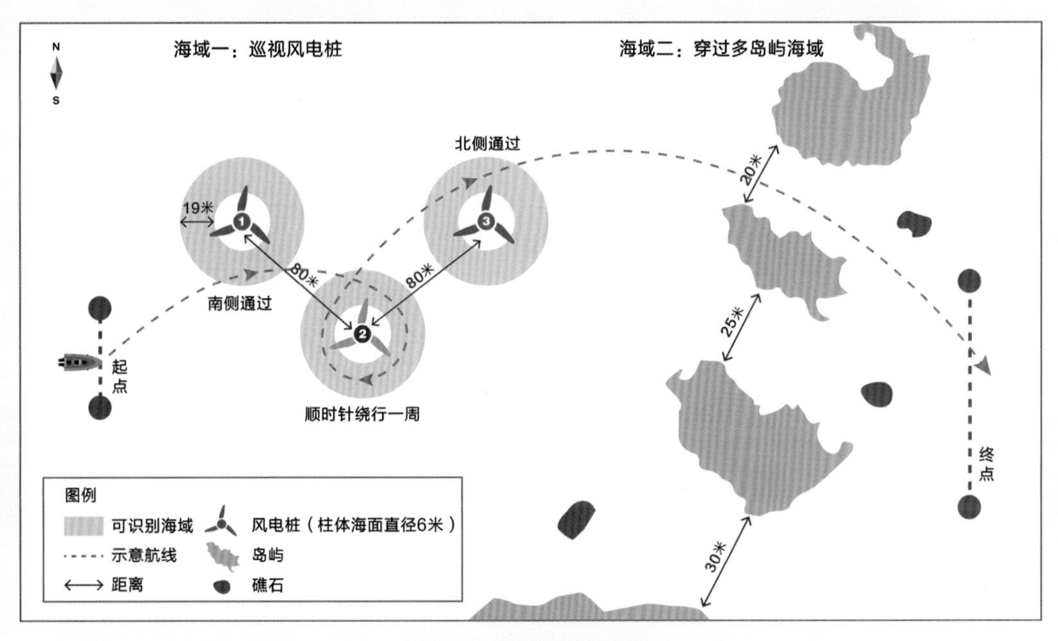


图1　初赛场景布置示意图

3号风电桩：北侧通过，过程中不得与风电桩碰撞且距离不得小于1米，绕行过程中与风电桩距离不能大于20米。

海域二：穿过多岛屿海域

参赛选手的无人艇采用自主选择线路航行的方式，从海域一继续往终点的方向航行，根据给定地图避开海中的岛屿与礁石，最终抵达终点线为完成比赛。

比赛要求如下。

（1）参赛选手需在3分钟内完成比赛，否则无成绩。

（2）船头通过起点线开始计时，船尾通过终点线时停止计时。

（3）通过风电桩海域时，无人艇与风电桩的距离不得小于1米，不得大于20米，若无人艇中点不在此范围内，则判定为任务未完成。

（4）海况风速为0~3级，流速为0~0.2米/秒。

（5）每队有两次比赛机会，任一次完成任务即可。

2）决赛规则

赛事预设场景范围为1200米×800米，要求在10分钟内巡检风电桩、穿过多岛屿海域并抵达终点。决赛按照分数进行最终排名，若得分相同，则用时较短者获胜。决赛场景布置示意图（图2）中海域一场景与决赛实际场景一致，海域二场景布置仅供参考，决赛现

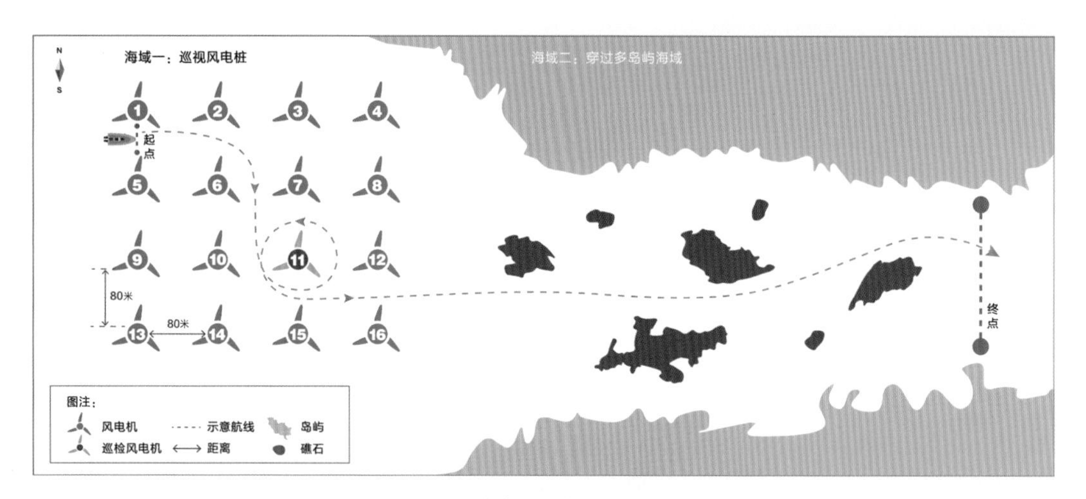


图2　决赛场景布置示意图

场地图赛前确定，请仔细阅读以下规则。

注：由于赛事需要，场景中距离参数相应缩短。技术参数初赛后发布。

海域一：巡检风电桩

穿越4×4风电桩阵列，每两个相邻风电桩的距离为80米。参赛选手的无人艇从1号和5号风电桩之间的起点出发，从起点往终点的方向，采用自主选择线路航行的方式，完成对11号风电桩绕行一周的巡检任务（巡检无动作），并从12号和16号风电桩之间穿出。

海域二：穿过多岛屿海域

参赛选手的无人艇从12号和16号风电桩之间穿出后继续航行，经由狭长多岛屿海域，自主选择线路航行并抵达终点。航行过程中需要避开海域中的岛屿与礁石，水道的最小宽度为15米，不提供岛屿和礁石的坐标位置（图2中水道区域非最终决赛场景）。

计分规则如下。

（1）船头通过起点线开始计时，船尾到达终点线停止计时。

（2）在海域一完成风电桩巡检任务得45分。其中包括正确从1号和5号之间出发得10分，从12号和16号之间离开风电桩阵列得10分，围绕11号风电桩巡检一周得25分。完成巡检任务过程中，若无人艇与各风电桩的距离小于1米，则每次扣5分；若撞击风电桩，则每次扣10分；扣分不超过该段总得分。

（3）在海域二通过多岛屿海域，成功到达终点线得45分。若无人艇与礁石或岛屿相碰撞，则每次减5分；扣分不超过该段总得分。

（4）规定时间内完成比赛得10分。决赛时间为10分钟，每超时1分钟扣5分，总赛程超过12分钟自动终止比赛。

（5）决赛满分100分。

（6）每支参赛队伍只有一次比赛机会。

注：本规则的解释权归大赛组委会所有。

## 1.Introduction

With the depletion of traditional energy sources and the maturity of wind power generation technology, wind power generation has become an important source of new energy power generation worldwide. China's offshore exploitable wind energy resources are about 750 million kilowatts, which is three times that of onshore wind energy resources. In recent years, the development of China's offshore wind power industry is quite rapid, with increasing offshore wind farm numbers and scales. However, due to differences in the marine and land environments, the rapid development of offshore wind farms has also posed challenges to daily inspection and maintenance operations. Therefore, in order to reduce economic costs and improve operation and maintenance efficiency, the competition conducts simulation propositions around the offshore wind power patrol navigation routes. To programme on the offshore wind power operation and maintenance simulation platform (please download the platform on the competition website), and complete the tasks of marine inspection and autonomous obstacle avoidance under different sea conditions with designated movements within the allotted time.

## 2.Form

Virtual simulation navigation.

## 3.Participants

University group: graduate students, undergraduates, and junior college students;

There are no more than 4 players (including the team leader) in each group.

## 4.Rules

The simulation preliminary and final contests have different scenarios. The preliminary format is a time-limited task, and the final is based on a score ranking system. Python, C++, C#, Java are recognized as programming language for this competition.

1) Preliminary contest rules

The preset scenario scale is 800×600 meters. Please patrol the wind power piles as required, and pass through multi-island sea areas, then reach the finishing line in total three minutes. The schematic map below is the real scenario of the preliminary contest, please read through the following rules carefully.

Note: As the competition needed, the distance parameter in the scenario is shortened accordingly. The technical parameters are detailed in the simulation platform operation manual.

Sea area I: to patrol the wind power piles

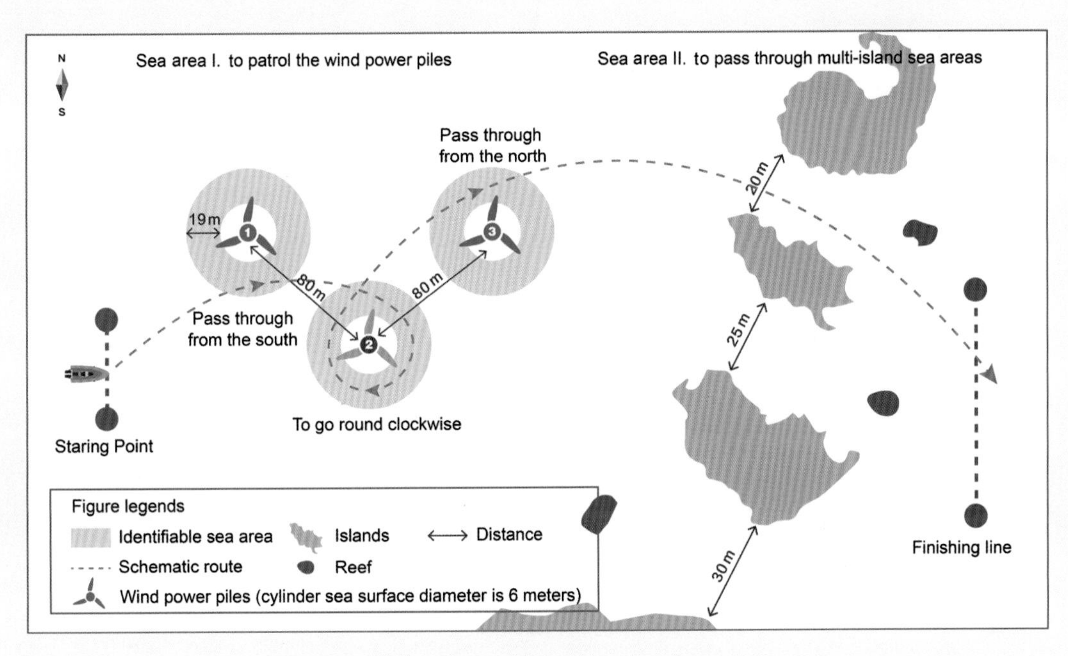


Figure 1 Preliminary contest scenario schematic map

The unmanned boats start from the starting point, use autonomous sailing to successively arrive at the identifiable sea area (blue area) near each wind pile, and complete required movements, then continue to sail towards the finishing line. The required movements of each wind power pile are as follows:

No. 1 wind power pile: to pass from the south without bumping into the pile, and to keep a distance of no less than 1 meter from it. During the bypass, the distance from the wind power pile shall not exceed 20 meters.

No. 2 wind power pile: to conduct one round of clockwise inspection in the blue identifiable sea area without bumping into the pile, and to keep a distance of no less than 1 meter from it. During the bypass, the distance from the wind power pile shall not exceed 20 meters.

No. 3 wind power pile: to pass from the north side without bumping into the pile, and to keep a distance of no less than 1 meter from it. During the bypass, the distance from the wind power pile shall not exceed 20 meters.

Sea area II: to pass through multi-island sea areas

Unmanned boats independently select routes to sail towards the finishing line, avoiding islands and reefs in the sea according to the schematic map, and reaching the finishing line as the completion of the competition.

The requirements are as follows:

(1) Contestants must complete the competition within three minutes; otherwise, there will be no scores.

(2) Timekeeping starts when the bow reaches the starting line, and stops when the stern crosses the finishing line.

(3) When passing through the wind power pile sea area, the distance between the unmanned boat and the pile shall not be less than 1 meter, not more than 20 meters. And if the midpoint of the unmanned boat is not within this distance range, the task will be judged as unfinished.

(4) The sea conditions, wind speed is 0~3 level, and the flow rate is 0~0.2 m/s.

(5) Each team has two chances, and once completion is regarded as finishing the preliminary contest.

2) Final rules

The preset scenario scale is 1200×800 meters. Please patrol the wind power piles as required, and pass through multi-island sea areas, then reach the finishing line in total 10 minutes. The final is based on a score ranking system, and if scores are the same, the team that uses the shortest time wins. The sea area I in the schematic map below is the same as the real scenario of the final, while the Sea Area II is for reference only, and the map will be generated before the final. Please read through the following rules carefully.

Note: As the competition needed, the distance parameter in the scenario is shortened accordingly. The technical parameters will be released after the preliminary contest.

Sea area I: to patrol the wind power piles

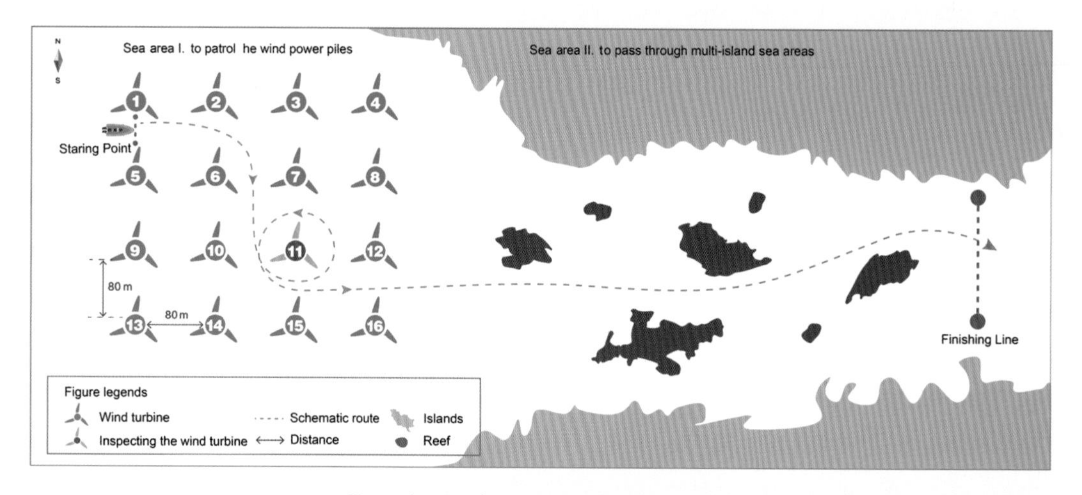


Figure2 Final scenario schematic map

To cross 4×4 wind power pile arrays, each adjacent wind power pile with a distance of 80 meters. The unmanned boats start from the starting point between No. 1 and No. 5 wind power pile, towards the finishing line. To independently select routes to complete a round inspection of the No. 11 wind power pile (no required movement), and pass through between the No. 12 and No. 16 wind power pile.

Sea area II: to pass through multi-island sea areas

After passing through between the No. 12 and No. 16 wind power pile, unmanned boats continue to sail through the narrow multi-island waters, and then independently select routes towards the finishing line. During the voyage, they should avoid islands and reefs in the sea, the minimum width of the waterway is 15 meters, and position coordinates of the islands and reefs are not provided (the waterway area in the schematic map is not the final scenario).

The final scoring rules are as follows:

(1) Timekeeping starts when the bow reaches the starting line, and stops when the stern crosses the finishing line.

(2) The total score for completing the sea area I patrol is 45, including 10 for correct departure between No. 1 and No. 5 wind power pile, another 10 for leaving the wind power pile arrays between the No. 12 and No. 16 wind power pile, and 25 for round routing inspection of No. 11. In the process of the inspection, if the distance between the boat and each wind power pile is less than 1 meter, 5 scores will be deducted each time. If the boat hit the wind power pile, 10 scores will be deducted each time. And the deduction will not exceed the total score of the section.

(3) The total score for passing the sea area II and reaching the finishing line is 45. If the boat strikes reef or island, 5 scores will be deducted each time, and the deduction will not exceed the total score of the section.

(4) The time limit is 10 minutes, and 10 scores for completing on time. 5 scores will be deducted for each overtime minute, and the schedule exceeds 12 minutes to automatically terminate the competition.

(5) The full score is 100.

(6) Each team has only one chance in the final.

Note: The organizing committee reserves the right of final interpretation.